T0400777

SCIENTIFIC REVOLUTIONS

ARTIFICIAL INTELLIGENCE: APPROACHES, TOOLS AND APPLICATIONS

SCIENTIFIC REVOLUTIONS

Additional books in this series can be found on Nova's website
under the Series tab.

Additional E-books in this series can be found on Nova's website
under the E-books tab.

COMPUTER SCIENCE, TECHNOLOGY AND APPLICATIONS

Additional books in this series can be found on Nova's website
under the Series tab.

Additional E-books in this series can be found on Nova's website
under the E-books tab.

SCIENTIFIC REVOLUTIONS

ARTIFICIAL INTELLIGENCE: APPROACHES, TOOLS AND APPLICATIONS

BRENT M. GORDON
EDITOR

Nova Science Publishers, Inc.
New York

For permission to use material from this book please contact us:
Telephone 631-231-7269; Fax 631-231-8175
Web Site: http://www.novapublishers.com

Library of Congress Cataloging-in-Publication Data

Artificial intelligence : approaches, tools, and applications / [edited by] Brent M. Gordon.
p. cm.
Includes index.
ISBN 978-1-61324-019-9 (hardcover)
1. Artificial intelligence. I. Gordon, Brent M.
Q335.5.A783 2011
006.3--dc22
2011006701

Published by Nova Science Publishers, Inc. † New York

CONTENTS

PREFACE

Artificial Intelligence may be defined as a collection of several analytic tools that collectively attempt to imitate life and has matured to a set of analytic tools that facilitate solving problems which were previously difficult or impossible to solve. In this new book, the authors present topical research in the study of the tools and applications of artificial intelligence. Topics discussed include the application of artificial intelligence in the oil and gas industry and in metal stamping die design; using artificial intelligence to predict embryo quality and in biomedical imaging techniques. (Imprint: Nova Press).

Chapter 1 – This chapter starts with an introduction to Artificial Intelligence (AI) including some historical background of the technology. Artificial Intelligence may be defined as a collection of several analytic tools that collectively attempt to imitate life.[1] In the last twenty years Artificial Intelligence has matured to a set of analytic tools that facilitate solving problems which were previously difficult or impossible to solve. The trend now is the integration of these tools, as well as with conventional technologies such as statistical analysis, to build sophisticated systems capable of solving challenging problems. Artificial Intelligence is used in areas such as medical diagnosis, credit card fraud detection, bank loan approval, smart household appliances, subway systems, automatic transmissions, financial portfolio management, robot navigation systems, and many more. In the oil and gas industry these tools have been used to solve problems related to pressure transient analysis, well log interpretation, reservoir characterization, and candidate well selection for stimulation, among others. Artificial neural networks, evolutionary programming and fuzzy logic are among the paradigms that are classified as Artificial Intelligence. These technologies exhibit an ability to learn and deal with new situations by possessing one or more attributes of "reason", such as generalization, discovery, association and abstraction [2]. This chapter is organized in four sections. First three sections are focused on the definition of some of the AI tools which are most commonly practiced and show some of their applications in the upstream oil and gas industry. In the first section artificial neural networks, are introduced as information processing systems that have certain performance characteristics in common with biological neural networks. This section will conclude with detail presentation of an application of neural networks in the upstream oil and gas industry. Second section is intended to provide an overview of evolutionary computing, its potential combination with neural networks to produce powerful intelligent applications, and its applications in the oil and gas industry. The most successful intelligent applications incorporate several artificial intelligence tools in a hybrid manner. These tools complement each other and amplify each other's effectiveness. An overview of evolutionary computation and its background is presented, followed by a more detailed look at genetic algorithms as the primary evolutionary computing. The article will continue and conclude by exploring the application of a hybrid neural network/genetic algorithm system to a petroleum engineering related problem. Fuzzy logic is the focus of the third section. An overview of the subject is provided followed by its potential application in petroleum engineering related problems. In this section, application of fuzzy logic for re-stimulation candidate selection in a tight gas formation in the Rocky Mountains will be reviewed. This particular application was chosen because it uses fuzzy logic in a hybrid manner integrated with neural networks and genetic algorithms. In the fourth and final section, some other applications of these tools in reservoir characterization, production modeling and performance prediction are presented.

Chapter 2 – One of the most recent applications of laser marking process is in the manufacturing of decorative gold. Gloss of the final gold marked is a criterion to evaluate the quality of product in terms of aesthetics appearance. This property essentially affected by various laser marking parameters such as laser power, feed rate (speed), Q-switch frequency (QSF) and pulse width (PW). In this paper, an adaptive neuro-fuzzy inference system (ANFIS) technique and artificial neural networks (ANNs) were utilized to model the effect of the mentioned parameters on the gloss of the laser marked gold. Both models were trained with experimental data. The results of this part of study indicated that ANNs had better outcomes compared to ANFIS. The best model was a cascade-forward backpropagation(CFBP) network, with various threshold functions (TANSIG-TANSIG-LOGSIG) and 9/8 neurons in the first/second hidden layers. Afterwards, in order to find the mentioned parameters of laser marking process,which maximize the gloss of the gold, the genetic algorithm (GA) and particle swarm optimization (PSO)were utilized and the best model was presented to the GA and PSO as the objective function. After the optimization, results of this part revealed that GA had better outcome compared to PSO so thatthe calculated gloss effect increases by 15% and the measured value increases by 12% in an experiment as compared to a non-optimized case.

Chapter 3 – Metal stamping die design is a tedious, time-consuming and highly experience based activity. Various artificial intelligence (AI) techniques are being used by worldwide researchers for stamping die design to reduce complexity, dependence on human expertise and time taken in design process as well as to improve design efficiency. In this chapter, various sheet metal operations, types of press tools, and AI techniques are briefly discussed. Further, a comprehensive review of applications of AI techniques to metal stamping die design is presented. The salient features of major research work published in the area of metal stamping are presented in tabular form. Thereafter, procedure for development of a knowledge-based system (KBS) for intelligent design of press tools is described at length. An intelligent system developed for quick design of progressive press tool is also presented. Finally, scope of future research work is identified.

Chapter 4 – In this study, structural features ofAlumina-Titanium diboridenanocomposite (Al_2O_3-TiB_2) were simulated from the mixture of titanium dioxide, boric acid and pure aluminum as raw materials via mechanochemical process using Artificial Intelligence approaches (AI). The phase transformation and structural evolutions during the mechanochemical process were characterized using X-Ray powder Diffractometry(XRD). For better understanding the refining crystallite size and amorphization phenomenaduring the milling, XRD data were modeled using Adaptive Neuro Fuzzy Inference System (ANFIS) and Artificial Neural Networks (ANN). Results show that ANN has better performance compared to ANFIS. The best predictor is then selected for simulation of crystallite size, interplaner distance, amorphization degree, and lattice strain. Furthermore, the simulated results are compared with experimental results.A good agreementbetween the experimental results and simulation ones were achieved.

Chapter 5 – One of the most relevant aspects in Assisted Reproductive Technologies is the characterization of the embryos to transfer in a patient. Objective assessment of embryo quality is actually an important matter of investigation both for bioethical and economical reasons. In most cases, embryologists evaluate embryos by visual examination and their evaluation is totally subjective. Recently, due to the rapid growth in our capacity to extract texture descriptors from a given image, a growing interest has been shown on the study of artificial intelligence methods to improve success rates of IVF programs based on the analysis and selection of images of embryos or oocytes. In this work we concentrate our efforts on the automatic classification of the quality of an embryo starting from the analysis of its image. The artificial intelligence system proposed in this work is based on textural descriptors (i.e. features), used to characterize the embryos, by measuring the homogeneity of their texture and the presence of recurrent patterns. A general purpose classifier is trained using visual descriptors to score the embryo images. The proposed system is tested on a datasets of 257 images with valuable classification results.

Chapter 6 – The new kind of probabilistic safety assessment (PSA) method has been studied for the very high temperature reactor (VHTR) which is a type of nuclear power plants (NPPs). There is a difficulty to make the quantification of the PSA, because the operation and experience data are deficient. Hence, it is necessary to manipulate the data statistically in basic events. The non-linear fuzzy set algorithm is used to quantification of the designed case for the physical data. The mass flow rate of the natural circulation is a main model. In addition, the potential energy in the gravity, the temperature and pressure in the heat conduction, and the heat transfer rate in the internal stored energy are also investigated. The values in the probability set and the fuzzy set are compared for the failure analysis. The results show the failure frequency in the propagations. It is concluded the artificial intelligence analysis of the fuzzy set could enhance the reliability than the probabilistic analysis. Namely, the dynamical safety assessment for the rare events has been newly developed. It is analyzed for the non-linear algorithm to substitute the probabilistic descriptions. The characteristics of the probabilistic distribution like the mean value and the standard deviation are changed to the some geometrical configuration as the slope and the radius in the fuzzy distribution. So, the operator can express the conceptions of the physical variable much more easily and exactly. The meaning of the non-linear algorithm shows the priority of the analysis. Using interpretation of the fuzzy set distributions, the quantity of the physical variables can be showed with the linguistic expression of the operator. Therefore, for the further study, the human error could be reduced due to the human oriented algorithm of the theory in some active systems, because the fuzzy set theory is originated from the linguistic expression of the operator. In addition, the other complex algorithm like the neural network or chaos theory could be applied to the data quantification in PSA.

Chapter 7 – The historical origin of the Artificial Intelligence (AI) is usually established in the Darmouth Conference, of 1956. But we can find many more arcane origins [1]. Also, we can consider, in more recent times, very great thinkers, as Janos Neumann (then, John von Neumann, arrived in USA), Norbert Wiener, Alan Mathison Turing, or Lofti Zadeh, for instance [12, 14]. Frequently AI requires Logic. But its Classical version shows too many insufficiencies. So, it was necessary to introduce more sophisticated tools, as Fuzzy Logic, Modal Logic, Non-Monotonic Logic and so on [1, 2]. And we need a new Mathematics, more adapted to the real world, which may comprises new areas, as Fuzzy Sets, Rough Sets, and Hybrid Systems; for instance, analyzing Knowledge with Uncertainty. Among the things that AI needs to represent are *Categories*, *Objects*, *Properties*, *Relations between objects*, *Situations*, *States*, *Time*, *Events*, *Causes and effects*, *Knowledge about knowledge*, *and so on.* The problems in AI can be classified in two general types [3, 5], *Search Problems* and *Representation Problems.* On this last "peak", there exist different ways to reach their summit. So, we have [4] *Logics, Rules, Frames, Associative Nets, Scripts*, and so on, many times connected among them. We attempt, in this paper, a panoramic vision of the scope of application of such emergent methods in AI.

Chapter 8 – In this chapter, we review advances of Artificial Intelligence (AI) applications, namely Artificial Neural Networks (ANNs), for the micro-Computed Tomography, a biomedical imaging technique. AI, and particularly ANNs, have yielded outstanding results in biomedical imaging, especially in image processing, feature extraction, classification and image interpretation. ANNs become a great ally for biomedical image analyses, in cases which traditional imaging approaches are not sufficient to detect specific characteristics, intrinsic patterns, or when computer-aided diagnosis must be sensitive to details associated to certain level of perception in the visualization. We discuss concepts related to this application of ANNs (training strategies), presenting results of the successful use of this technique.

In: Artificial Intelligence
Editor: Brent M. Gordon, pp. 1-38

ISBN 978-1-61324-019-9

Chapter 1

APPLICATION OF ARTIFICIAL INTELLIGENCE IN THE UPSTREAM OIL AND GAS INDUSTRY

Shahab D. Mohaghegh and Yasaman Khazaeni
Intelligent Solutions, Inc., and West Virginia University, Morgantown, West Virginia, USA

ABSTRACT

This chapter starts with an introduction to Artificial Intelligence (AI) including some historical background of the technology. Artificial Intelligence may be defined as a collection of several analytic tools that collectively attempt to imitate life.[1] In the last twenty years Artificial Intelligence has matured to a set of analytic tools that facilitate solving problems which were previously difficult or impossible to solve. The trend now is the integration of these tools, as well as with conventional technologies such as statistical analysis, to build sophisticated systems capable of solving challenging problems.

Artificial Intelligence is used in areas such as medical diagnosis, credit card fraud detection, bank loan approval, smart household appliances, subway systems, automatic transmissions, financial portfolio management, robot navigation systems, and many more. In the oil and gas industry these tools have been used to solve problems related to pressure transient analysis, well log interpretation, reservoir characterization, and candidate well selection for stimulation, among others.

Artificial neural networks, evolutionary programming and fuzzy logic are among the paradigms that are classified as Artificial Intelligence. These technologies exhibit an ability to learn and deal with new situations by possessing one or more attributes of "reason", such as generalization, discovery, association and abstraction [2].

This chapter is organized in four sections. First three sections are focused on the definition of some of the AI tools which are most commonly practiced and show some of their applications in the upstream oil and gas industry. In the first section artificial neural networks, are introduced as information processing systems that have certain performance characteristics in common with biological neural networks. This section will conclude with detail presentation of an application of neural networks in the upstream oil and gas industry.

Second section is intended to provide an overview of evolutionary computing, its potential combination with neural networks to produce powerful intelligent applications, and its applications in the oil and gas industry. The most successful intelligent applications incorporate several artificial intelligence tools in a hybrid manner. These tools complement each other and amplify each other's effectiveness. An overview of evolutionary computation and its background is presented, followed by a more detailed look at genetic algorithms as the primary evolutionary computing. The article will continue and conclude by exploring the application of a hybrid neural network/genetic algorithm system to a petroleum engineering related problem.

Fuzzy logic is the focus of the third section. An overview of the subject is provided followed by its potential application in petroleum engineering related problems. In this section, application of fuzzy logic for re-stimulation candidate selection in a tight gas formation in the Rocky Mountains will be reviewed. This particular application was chosen because it uses fuzzy logic in a hybrid manner integrated with neural networks and genetic algorithms.

In the fourth and final section, some other applications of these tools in reservoir characterization, production modeling and performance prediction are presented.

1. Neural Networks and Their Background

In this section some historical background of the technology will be mentioned followed by definitions of artificial intelligence and artificial neural networks. After the definitions, more general information on the nature and mechanism of the artificial neural network and its relevance to biological neural networks will be offered.

1.1. A Short History of Neural Networks

Neural network research can be traced back to a paper by McCulloch and Pitts [3] in 1943. In 1957 Frank Rosenblatt invented the Perceptron. [4] Rosenblatt proved that given linearly separable classes, a perceptron would, in a finite number of training trials, develop a weight vector that will separate the classes (a pattern classification task). He also showed that his proof holds independent of the starting value of the weights. Around the same time Widrow and Hoff [5] developed a similar network called Adeline. Minskey and Papert [6] in a book called "Perceptrons" pointed out that the theorem obviously applies to those problems that the structure is capable of computing. They showed that elementary calculation such as simple "exclusive or" (XOR) problems cannot be solved by single layer perceptrons.

Rosenblatt [4] had also studied structures with more layers and believed that they could overcome the limitations of simple perceptrons. However, there was no learning algorithm known which could determine the weights necessary to implement a given calculation. Minskey and Papert doubted that one could be found and recommended that other approaches to artificial intelligence should be pursued. Following this discussion, most of the computer science community left the neural network paradigm for twenty years [7]. In early 1980s Hopfield was able to revive the neural network research. Hopfield's efforts coincided with development of new learning algorithms such as backpropagation. The growth of neural network research and applications has been phenomenal since this revival.

1.2. Structure of a Neural Network

An artificial neural network is an information processing system that has certain performance characteristics in common with biological neural networks. Therefore it is appropriate to describe briefly a biological neural network before offering a detail definition of artificial neural networks. All living organisms are made up of cells. The basic building blocks of the nervous system are nerve cells, called neurons. Figure 1 shows a schematic diagram of two bipolar neurons.

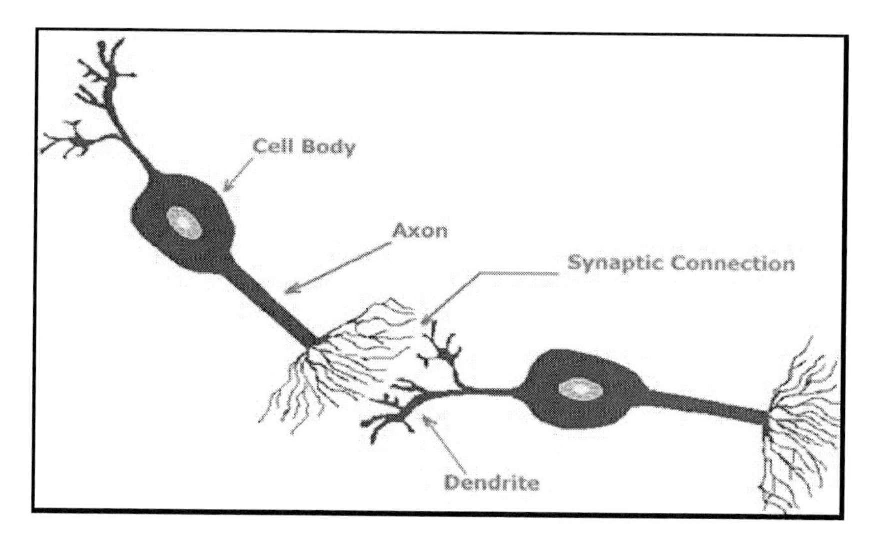

Figure 1. Schematic diagram of two bipolar neurons.

A typical neuron contains a cell body where the nucleus is located, dendrites and an axon. Information in the form of a train of electro-chemical pulses (signals) enters the cell body from the dendrites. Based on the nature of this input the neuron will activate in an excitatory or inhibitory fashion and provides an output that will travel through the axon and connects to other neurons where it becomes the input to the receiving neuron. The point between two neurons in a neural pathway, where the termination of the axon of one neuron comes into close proximity with the cell body or dendrites of another, is called a *synapse.* The signals traveling from the first neuron initiate a train of electro-chemical pulse (signals) in the second neuron.

It is estimated that the human brain contains on the order of 10 to 500 billion neurons [8]. These neurons are divided into modules and each module contains about 500 neural networks.[9] Each network may contain about 100,000 neurons in which each neuron is connected to hundreds to thousands of other neurons. This architecture is the main driving force behind the complex behavior that comes so natural to us. Simple tasks such as catching a ball, drinking a glass of water or walking in a crowded market require so many complex and coordinated calculations that sophisticated computers are unable to undertake the task, and yet is done routinely by humans without a moment of thought. This becomes even more interesting when one realizes that neurons in the human brain have cycle time of about 10 to 100 milliseconds while the cycle time of a typical desktop computer chip is measured in nanoseconds. The human brain, although million times slower than common desktop PCs, can perform many tasks orders of magnitude faster than computers because of it massively parallel architecture.

Artificial neural networks are a rough approximation and simplified simulation of the process explained above. An artificial neural network can be defined as an information processing system that has certain performance characteristics similar to biological neural networks. They have been developed as generalization of mathematical models of human cognition or neural biology, based on the assumptions that:

- Information processing occurs in many simple elements that are called neurons (processing elements).
- Signals are passed between neurons over connection links.
- Each connection link has an associated weight, which, in a typical neural network, multiplies the signal being transmitted.
- Each neuron applies an activation function (usually non-linear) to its net input to determine its output signal [10]

Figure 2 is a schematic diagram of a typical neuron (processing element) in an artificial neural network. Output from other neurons is multiplied by the weight of the connection and enters the neuron as input. Therefore an artificial neuron has many inputs and only one output. The inputs are summed and subsequently applied to the activation function and the result is the output of the neuron.

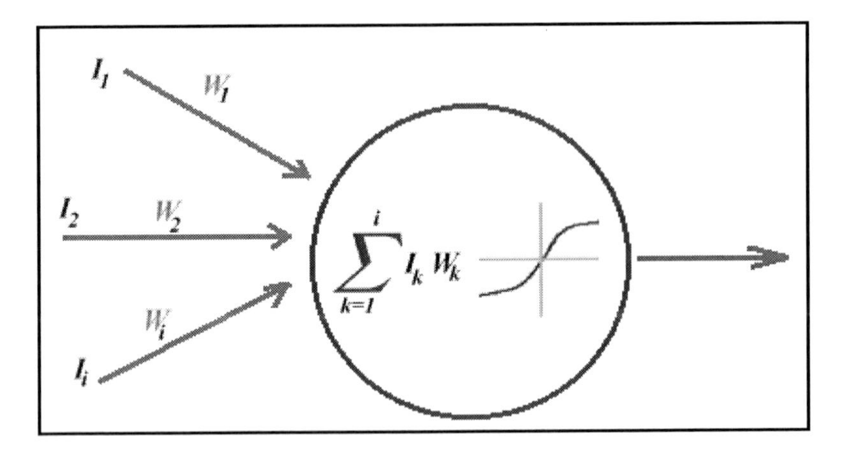

Figure 2. Schematic diagram of an artificial neuron or a processing element.

1.3. Mechanics of Neural Networks Operation

An artificial neural network is a collection of neurons that are arranged in specific formations. Neurons are grouped into layers. In a multi-layer network there are usually an input layer, one or more hidden layers and an output layer. The number of neurons in the input layer corresponds to the number of parameters that are being presented to the network as input. The same is true for the output layer. It should be noted that neural network analysis is not limited to a single output and that neural nets can be trained to build neuro-models with multiple outputs. The neurons in the hidden layer or layers are mainly responsible for feature extraction. They provide increased dimensionality and accommodate tasks such as classification and pattern recognition. Figure 3 is a schematic diagram of a fully connected three layered neural network.

There are many kinds of neural networks. Neural network scientists and practitioners have provided different classifications for neural networks. One of the most popular classifications is based on the training methods. Neural nets can be divided into two major categories based on the training methods, namely supervised and unsupervised neural networks. Unsupervised neural networks, also known as self-organizing maps, are mainly clustering and classification algorithms. They have been used in oil and gas industry to interpret well logs and to identify lithology. They are called unsupervised simply because no feedback is provided to the network. The network is asked to classify the input vectors into groups and clusters. This requires a certain degree of redundancy in the input data and hence the notion that redundancy is knowledge [11].

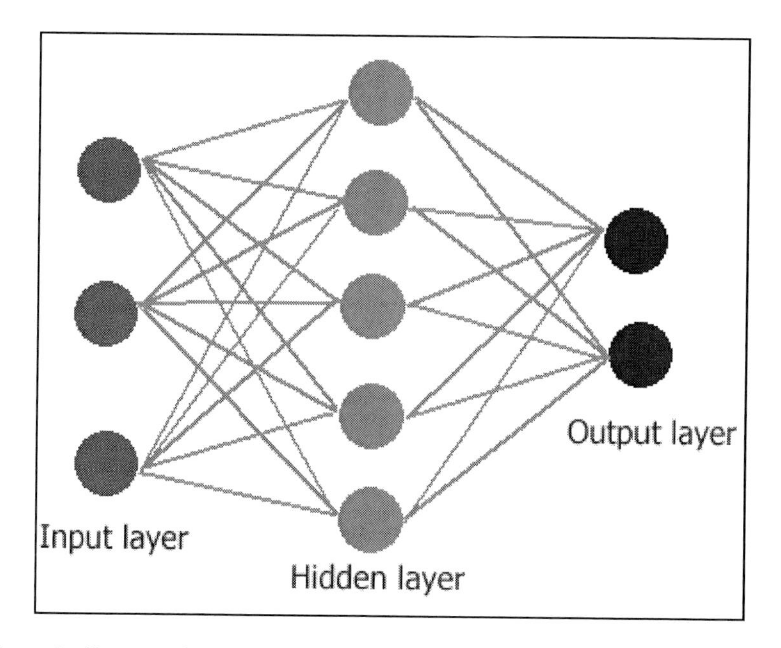

Figure 3. Schematic diagram of a three-layer neuron network.

Most of the neural network applications in the oil and gas industry are based on supervised training algorithms. During a supervised training process both input and output are presented to the network to permit learning on a feedback basis. A specific architecture, topology and training algorithm is selected and the network is trained until it converges. During the training process neural network tries to converge to an internal representation of the system behavior. Although by definition neural nets are model-free function approximators, some people choose to call the trained network a neuro-model.

The connections correspond roughly to the axons and synapses in a biological system, and they provide a signal transmission pathway between the nodes. Several layers can be interconnected. The layer that receives the inputs is called the input layer. It typically performs no function other than the buffering of the input signal. The network outputs are generated from the output layer. Any other layers are called hidden layers because they are internal to the network and have no direct contact with the external environment. Sometimes they are likened to a "black box" within the network system. However, just because they are not immediately visible does not mean that one cannot examine the function of those layers. There may be zero to several hidden layers. In a fully connected network every output from one layer is passed along to every node in the next layer.

In a typical neural data processing procedure, the database is divided into three separate portions called training, calibration and verification sets. The training set is used to develop the desired network. In this process (depending on the paradigm that is being used), the desired output in the training set is used to help the network adjust the weights between its neurons or processing elements. During the training process the question arises as when to stop the training. How many times should the network go through the data in the training set in order to learn the system behavior? When should the training stop? These are legitimate questions, since a network can be over trained. In the neural network related literature over-training is also referred to as memorization. Once the network memorizes a data set, it would be incapable of generalization. It will fit the training data set quite accurately, but suffers in generalization. Performance of an over-trained neural network is similar to a complex non-linear regression analysis.

Over-training does not apply to some neural network paradigms simply because they are not trained using an iterative process. Memorization and over-training is applicable to those networks that are historically among the most popular ones for engineering problem solving. These include back-propagation networks that use an iterative process during the training.

In order to avoid over training or memorization, it is a common practice to stop the training process every so often and apply the network to the calibration data set. Since the output of the calibration data set is not presented to the network, one can evaluate network's generalization capabilities by how well it predicts the calibration set's output. Once the training process is completed successfully, the network is applied to the verification data set.

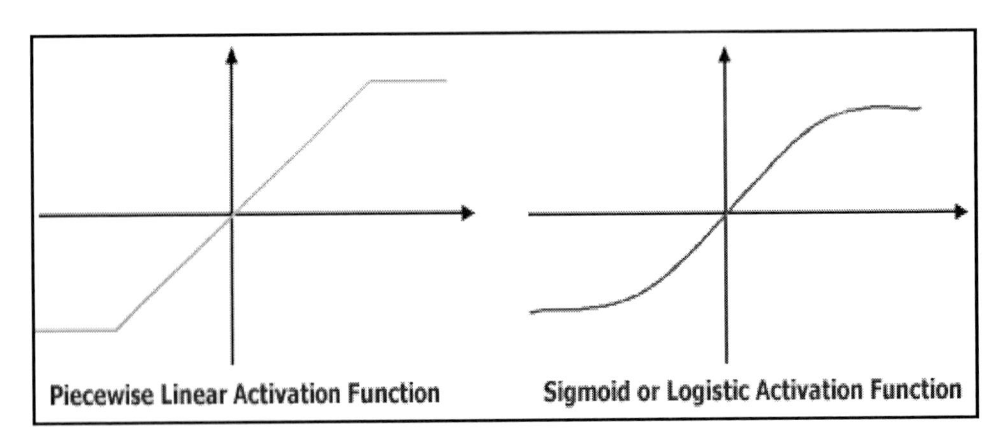

Figure 4. Commonly used activation functions in artificial neurons.

During the training process each artificial neuron (processing element) handles several basic functions. First, it evaluates input signals and determines the strength of each one. Second, it calculates a total for the combined input signals and compares that total to some threshold level. Finally, it determines what the output should be. The transformation of the input to output - within a neuron - takes place using an activation function. Figure 4 shows two of the commonly used activation (transfer) functions.

All the inputs come into a processing element simultaneously. In response, neuron either "fires" or "doesn't fire"; depending on some threshold level. The neuron will be allowed a single output signal, just as in a biological neuron - many inputs, one output. In addition, just as things other than inputs affect real neurons, some networks provide a mechanism for other influences. Sometimes this extra input is called a bias term, or a forcing term. It could also be a forgetting term, when a system needs to unlearn something [12].

Initially each input is assigned a random relative weight (in some advanced applications – based on the experience of the practitioner- the relative weight assigned initially may not be random). During the training process the weight of the inputs is adjusted. The weight of the input represents the strength of its connection to the neuron in the next layer. The weight of the connection will affect the impact and the influence of that input. This is similar to the varying synaptic strengths of biological neurons. Some inputs are more important than others in the way they combine to produce an impulse. Weights are adaptive coefficients within the network that determine the intensity of the input signal. The initial weight for a processing element could be modified in response to various inputs and according to the network's own rules for modification.

Mathematically, we could look at the inputs and the weights on the inputs as vectors, such as $I_1, I_2 \ldots I_n$ for inputs and $W_1, W_2 \ldots W_n$ for weights. The total input signal is the dot, or inner, product of the two vectors. Geometrically, the inner product of two vectors can be considered a measure of their similarity. The inner product is at its maximum if the vectors point in the same direction. If the vectors point in opposite directions (180 degrees), their inner product is at its minimum. Signals coming into a neuron can be positive (excitatory) or negative (inhibitory). A positive input promotes the firing of the processing element, whereas a negative input tends to keep the processing element from firing. During the training process some local memory can be attached to the processing element to store the results (weights) of previous computations. Training is accomplished by modification of the weights on a continuous basis until convergence is reached. The ability to change the weights allows the network to modify its behavior in response to its inputs, or to learn. For example, suppose a network identifies a production well as "an injection well." On successive iterations (training), connection weights that respond correctly to a production well are strengthened and those that respond to others, such as an injection well, is weakened until they fall below the threshold level and the correct recognition of the well is achieved.

In the back propagation algorithm (one of the most commonly used supervised training algorithms) the network output is compared with the desired output - which is part of the training data set, and the difference (error) is propagated backward through the network. During this back propagation of error the weights of the connections between neurons are adjusted. This process is continued in an iterative manner. The network converges when its output is within acceptable proximity of the desired output.

2. Evolutionary Computing

Evolutionary computing, like other virtual intelligence tools, has its roots in nature. It is an attempt to mimic the evolutionary process using computer algorithms and instructions. However, why would we want to mimic the evolution process? The answer will become obvious once we realize what type of problems the evolution process solves and whether we would like to solve similar problems. Evolution is an optimization process [13] One of the major principles of evolution is heredity. Each generation inherits the evolutionary characteristics of the previous generation and passes those same characteristics to the next generation. These characteristics include those of progress, growth and development. This passing of the characteristics from generation to generation is facilitated through genes.

Since the mid 1960s, a set of new analytical tools for intelligent optimization have surfaced that are inspired by the Darwinian evolution theory. The term "evolutionary computing" has been used as an umbrella for many of these tools. Evolutionary computing comprises of evolutionary programming, genetic algorithms, evolution strategies, and evolution programs, among others. For many people, these tools (and names) look similar and their names are associated with the same meaning. However, these names carry quite distinct meanings to the scientists deeply involved in this area of research. Evolutionary programming, introduced by John Koza [14], is mainly concerned with solving complex problems by evolving sophisticated computer programs from simple, task-specific computer programs. Genetic algorithms are the subject of this article and will be discussed in detail in the next section. In evolution strategies [15], the components of a trial solution are viewed as behavioral traits of an individual, not as genes along a chromosome, as implemented in genetic algorithms. Evolution programs [16] combine genetic algorithms with specific data structures to achieve its goals.

2.1. Genetic Algorithms

Darwin's theory of survival of the fittest (presented in his 1859 paper titled *On the Origin of Species by Means of Natural Selection*), coupled with the selectionism of Weismann and the genetics of Mendel, have formed the universally accepted set of arguments known as the evolution theory [15].

In nature, the evolutionary process occurs when the following four conditions are satisfied [14]:

- An entity has the ability to reproduce.
- There is a population of such self-reproducing entities.
- There is some variety among the self-reproducing entities.
- This variety is associated with some difference in ability to survive in the environment.

In nature, organisms evolve as they adapt to dynamic environments. The "fitness" of an organism is defined by the degree of its adaptation to its environment. The organism's fitness determines how long it will live and how much of a chance it has to pass on its genes to the next generation. In biological evolution, only the winners survive to continue the evolutionary process. It is assumed that if the organism lives by adapting to its environment, it must be doing something right. The characteristics of the organisms are coded in their genes, and they pass their genes to their offspring through the process of heredity. The fitter an individual, the higher is its chance to survive and hence reproduce.

Intelligence and evolution are intimately connected. Intelligence has been defined as the capability of a system to adapt its behavior to meet goals in a range of environments [15]. By imitating the evolution process using computer instructions and algorithms, researchers try to mimic the intelligence associated with the problem solving capabilities of the evolution process. As in real life, this type of continuous adaptation creates very robust organisms. The whole process continues through many "generations", with the best genes being handed down to future generations. The result is typically a very good solution to the problem. In computer simulation of the evolution process, genetic operators achieve the passing on of the genes from generation to generation. These operators (crossover, inversion, and mutation) are the primary tools for spawning a new generation of individuals from the fit individuals of the current population. By continually cycling these operators, we have a surprisingly powerful search engine. This inherently preserves the critical balance needed with an intelligent search: the balance between exploitation (taking advantage of information already obtained) and exploration (searching new areas). Although simplistic from a biologist's viewpoint, these algorithms are sufficiently complex to provide robust and powerful search mechanisms.

2.2. Mechanism of a Genetic Algorithm

The process of genetic optimization can be divided into the following steps:

1. Generation of the initial population.
2. Evaluation of the fitness of each individual in the population.
3. Ranking of individuals based on their fitness.
4. Selecting those individuals to produce the next generation based on their fitness.
5. Using genetic operations, such as crossover, inversion and mutation, to generate a new population.
6. Continue the process by going back to step 2 until the problem's objectives are satisfied.

The initial population is usually generated using a random process covering the entire problem space. This will ensure a wide variety in the gene pool. Each problem is encoded in the form of a chromosome. Each chromosome is collection of a set of genes. Each gene represents a parameter in the problem. In classical genetic algorithms, a string of 0s and 1s or a bit string represents each gene (parameter). Therefore, a chromosome is a long bit string that includes all the genes (parameters) for an individual. Figure 5 shows a typical chromosome as an individual in a population that has five genes. Obviously, this chromosome is for a problem that has been coded to find the optimum solution using five parameters.

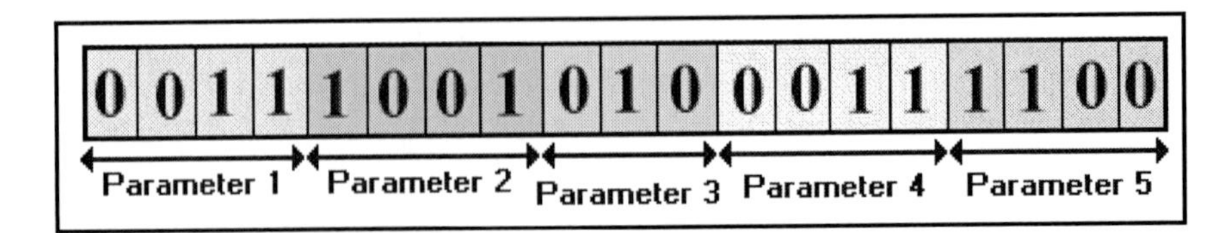

Figure 5. A chromosome with five genes.

The fitness of each individual is determined using a fitness function. The goal of optimization is usually to find a minimum or a maximum. Examples of this include the minimization of error for a problem that must converge to a target value or the maximization of the profit in a financial portfolio. Once the fitness of each individual in the population is evaluated, all the individuals will be ranked. After the ranking, it is time for selection of the parents that will produce the next generation of individuals. The selection process assigns a higher probability of reproduction to the highest-ranking individual, and the reproduction probability is reduced with a reduction in ranking.

After the selection process is complete, genetic operators such as crossover, inversion and mutation are incorporated to generate a new population. The evolutionary process of survival of the fittest takes place in the selection and reproduction stage. The higher the ranking of an individual is, the higher the chance for it to reproduce and pass on its gene to the next generation.

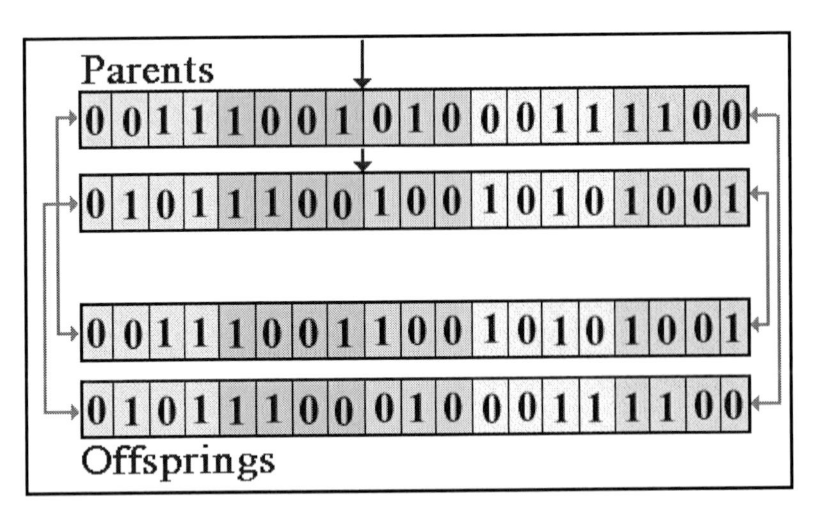

Figure 6. Simple crossover operator.

In crossover, the two parent individuals are first selected and then a break location in the chromosome is randomly identified. Both parents will break at that location and the halves switch places. This process produces two new individuals from the parents. One pair of parents may break in more than one location at different times to produce more than one pair of offspring. Figure 6 demonstrates the simple crossover.

There are other crossover schemes besides simple crossover, such as double crossover and random crossover. In double crossover, each parent breaks in two locations, and the sections are swapped. During a random crossover, parents may break in several locations. Figure 7 demonstrates a double crossover process.

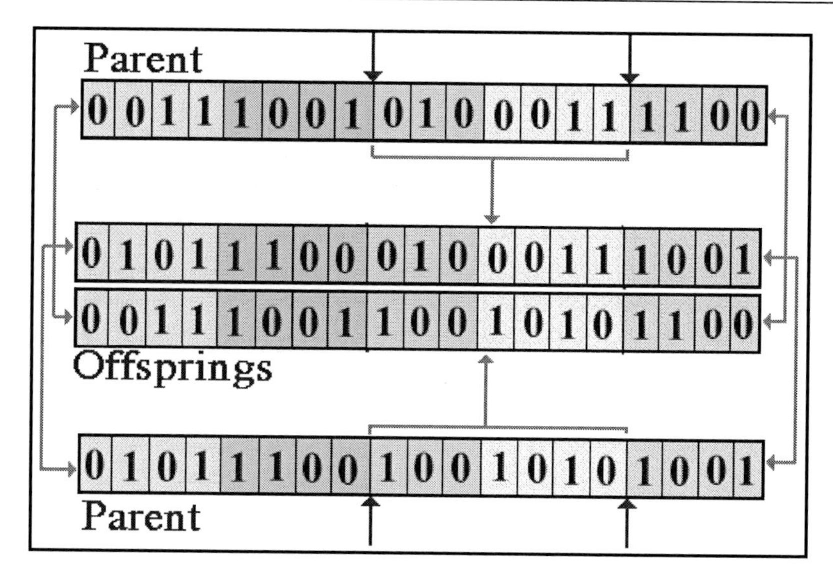

Figure 7. Double crossover operator.

As was mentioned earlier, there are two other genetic operators in addition to crossover. These are inversion and mutation. In both of these operators the offspring is reproduced from one parent rather than a pair of parents. The inversion operator changes all the 0s to 1s and all the 1s to 0s from the parent to make the offspring. The mutation operator chooses a random location in the bit string and changes that particular bit. The probability for inversion and mutation is usually lower than the probability for crossover. Figure 8 and Figure 9 demonstrate inversion and mutation.

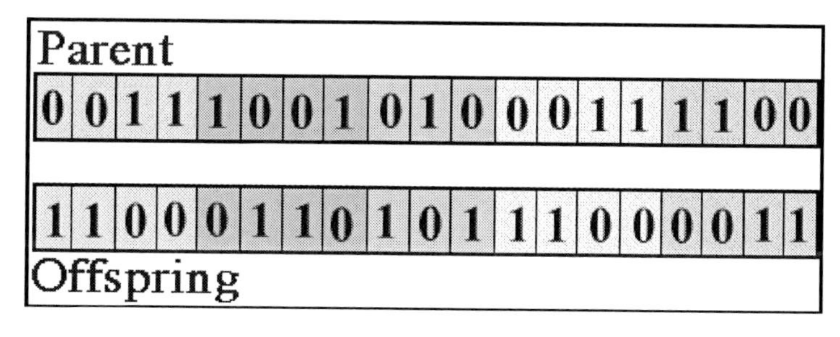

Figure 8. Inversion operator.

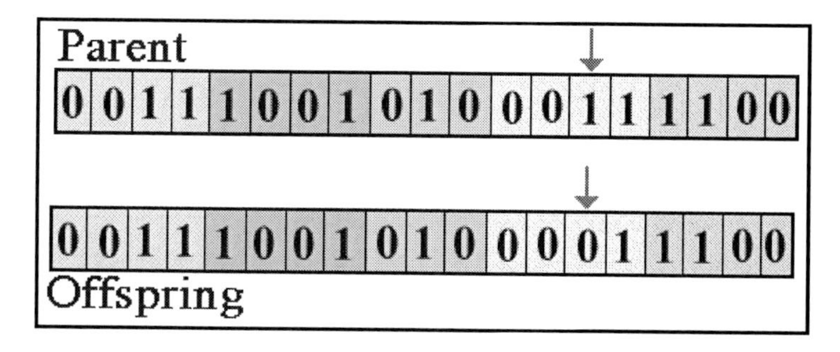

Figure 9. Mutation operator.

Once the new generation has been completed, the evaluation process using the fitness function is repeated and the steps in the aforementioned outline are followed. During each generation, the top ranking individual is saved as the optimum solution to the problem. Each time a new and better individual is evolved, it becomes the optimum solution. The convergence of the process can be evaluated using several criteria. If the objective is to minimize an error, then the convergence criteria can be the amount of error that the problem can tolerate. As another criterion, convergence can take place when a new and better individual is not evolved within four to five generations. Total fitness of each generation has also been used as a convergence criterion. Total fitness of each generation can be calculated (as a sum) and the operation can stop if that value does not improve in several generations. Many applications simply use a certain number of generations as the convergence criterion.

As you may have noticed, the above procedure is called the classic genetic algorithms. Many variations of this algorithm exist. For example, there are classes of problems that would respond better to genetic optimization if a data structure other than bit strings were used. Once the data structure that best fits the problem is identified, it is important to modify the genetic operators such that they accommodate the data structure. The genetic operators serve specific purposes – making sure that the offspring is a combination of parents in order to satisfy the principles of heredity – which should not be undermined when the data structure is altered.

Another important issue is introduction of constraints to the algorithm. In most cases, certain constraints must be encoded in the process so that the generated individuals are "legal". Legality of an individual is defined as its compliance with the problem constraints. For example in a genetic algorithm that was developed for the design of new cars, basic criteria, including the fact that all four tires must be on the ground, had to be met in order for the design to be considered legal. Although this seems to be quite trivial, it is the kind of knowledge that needs to be coded into the algorithm as constraints in order for the process to function as expected.

3. Fuzzy Logic

The science of today is based on Aristotle's crisp logic formed more than two thousand years ago. The Aristotelian logic looks at the world in a bivalent manner, such as black and white, yes and no, and 0 and 1. Development of the set theory in the late 19th century by German mathematician George Cantor that was based on the Aristotle's bivalent logic made this logic accessible to modern science. Then, the subsequent superimposition of probability theory made the bivalent logic reasonable and workable. Cantor's theory defines sets as a collection of definite, distinguishable objects. Figure 10 is a simple example of Cantor's set theory and its most common operations such as complement, intersection and union.

First work on vagueness dates back to the first decade of 1900, when American philosopher Charles Sanders Peirce noted that "vagueness is no more to be done away with in the world of logic than friction in mechanics [17]." In the early 1920s, Polish mathematician and logician Jan Lukasiewicz came up with three-valued logic and talked about many-valued or multi-valued logic [18]. In 1937, quantum philosopher Max Black published a paper on vague sets [19](19)(Black, 1937). These scientists built the foundation upon which fuzzy logic was later developed.

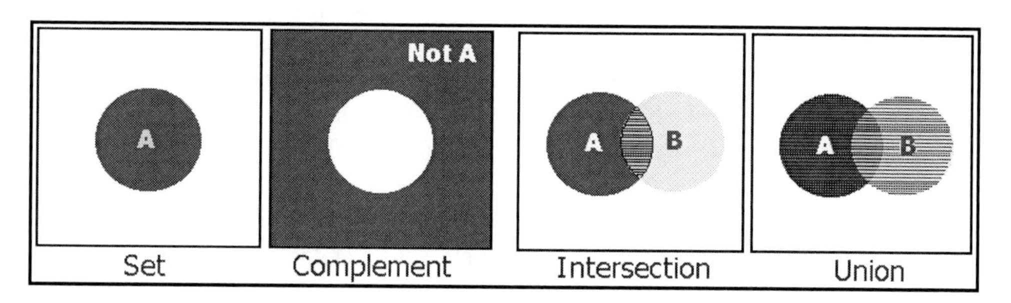

Figure 10. Operations of conventional crisp sets.

Lotfi A. Zadeh is known to be the father of fuzzy logic. In 1965, while he was the chair of the electrical engineering department at UC Berkeley, he published his landmark paper “Fuzzy Sets” [20]. Zadeh developed many key concepts including the membership values and provided a comprehensive framework to apply the theory to many engineering and scientific problems. This framework included the classical operations for fuzzy sets, which comprises all the mathematical tools necessary to apply the fuzzy set theory to real world problems. Zadeh used the term “fuzzy” for the first time, and with that he provoked much opposition. He became a tireless spokesperson for the field. He was often harshly criticized. For example, Professor R. E. Kalman said in a 1972 conference in Bordeaux, “Fuzzification is a kind of scientific permissiveness; it tends to result in socially appealing slogans unaccompanied by the discipline of hard scientific work [2].” (It should be noted that Kalman is a former student of Zadeh’s and the inventor of famous Kalman filter, a major statistical tool in electrical engineering. Kalman filter is the technology behind the Patriot missiles used in the Gulf War to shoot down Iraqi SCUD missiles. There have been claims that it has been proven that use of fuzzy logic can increase the accuracy of the Patriot missiles considerably.[21, 22] Despite all the adversities fuzzy logic continued to flourish and has become a major force behind many advances in intelligent systems.

The term “fuzzy” carries a negative connotation in the western culture. The term “fuzzy logic” seems to both misdirect the attention and to celebrate mental fog [23]. On the other hand, eastern culture embraces the concept of coexistence of contradictions as it appears in the Yin-Yang symbol. While Aristotelian logic preaches A or Not-A, Buddhism is all about A and Not-A.

Figure 11. The Yin-Yang symbol.

Many believe that the tolerance of eastern culture for such ideas was the main reason behind the success of fuzzy logic in Japan. While fuzzy logic was being attacked in the United States, Japanese industries were busy building a multi-billion dollar industry around it. Today, Japanese hold more than 2000 fuzzy related patents. They have used the fuzzy technology to build intelligent household appliances such as washing machines and vacuum cleaners (Matsushita and Hitachi), rice cookers (Matsushita and Sanyo), air conditioners (Mitsubishi), and microwave ovens (Sharp, Sanyo, and Toshiba), to name a few. Matsushita used fuzzy technology to develop its digital image stabilizer for camcorders. Adaptive fuzzy systems (a hybrid with neural networks) can be found in many Japanese cars. Nissan has patented a fuzzy automatic transmission that is now very popular with many other cars such as Mitsubishi and Honda [23].

3.1. Fuzzy Set Theory

The human thought, reasoning, and decision-making process is not crisp. We use vague and imprecise words to explain our thoughts or communicate with one another. There is a contradiction between the imprecise and vague process of human reasoning, thinking, and decision-making and the crisp, scientific reasoning of black and white computer algorithms and approaches. This contradiction has given rise to an impractical approach of using computers to assist humans in the decision-making process, which has been the main reason behind the lack of success for traditional artificial intelligence or conventional rule-based systems, also known as expert systems. Expert systems as a technology started in early 1950s and remained in the research laboratories and never broke through to consumer market.

In essence, fuzzy logic provides the means to compute with words. Using fuzzy logic, experts no longer are forced to summarize their knowledge to a language that machines or computers can understand. What traditional expert systems failed to achieve finally became reality (as mentioned above) with the use of fuzzy expert systems. Fuzzy logic comprises of fuzzy sets, which are a way of representing non-statistical uncertainty and approximate reasoning, which includes the operations used to make inferences [2].

Fuzzy set theory provides a means for representing uncertainty. Uncertainty is usually either due to the random nature of events or due to imprecision and ambiguity of information we have about the problem we are trying to solve. In a random process, the outcome of an event from among several possibilities is strictly the result of chance. When the uncertainty is a product of randomness of events, probability theory is the proper tool to use. Observations and measurements can be used to resolve statistical or random uncertainty. For example, once a coin is tossed, no more random or statistical uncertainty remains.

Most uncertainties, especially when dealing with complex systems, are the result of a lack of information. The kind of uncertainty that is the outcome of the complexity of a system is the type of uncertainty that rises from imprecision, from our inability to perform adequate measurements, from a lack of knowledge, or from vagueness (like the fuzziness inherent in natural language). Fuzzy set theory is a marvelous tool for modeling the kind of uncertainty associated with vagueness, with imprecision, and/or with a lack of information regarding a particular element of the problem at hand.[24] Fuzzy logic achieves this important task through fuzzy sets. In crisp sets, an object either belongs to a set or it does not. In fuzzy sets, everything is a matter of degrees. Therefore, an object belongs to a set to a certain degree. For example, the price of oil today is \$24.30 per barrel. Given the price of oil in the past few years, this price seems to be high. But what is a high price for oil? A few months ago, the price of oil was about \$10.00 per barrel. Everybody agrees that \$10.00 per barrel is low. Given how much it costs to produce a barrel of oil in the United States, one can say that the cut-off between low and high for oil price is \$15.00 per barrel. If we use crisp sets, then \$14.99 is low, and \$15.01 is high. However, imagine if this was the criterion that was used by oil company executives to make a decision. The fact is, while \$15.01 is a good price that many people will be happy with, \$16.00 is better, and \$20.00 is even better. Categorizing all these prices as high can be quite misleading. Fuzzy logic proposes the following fuzzy sets for the price of oil.

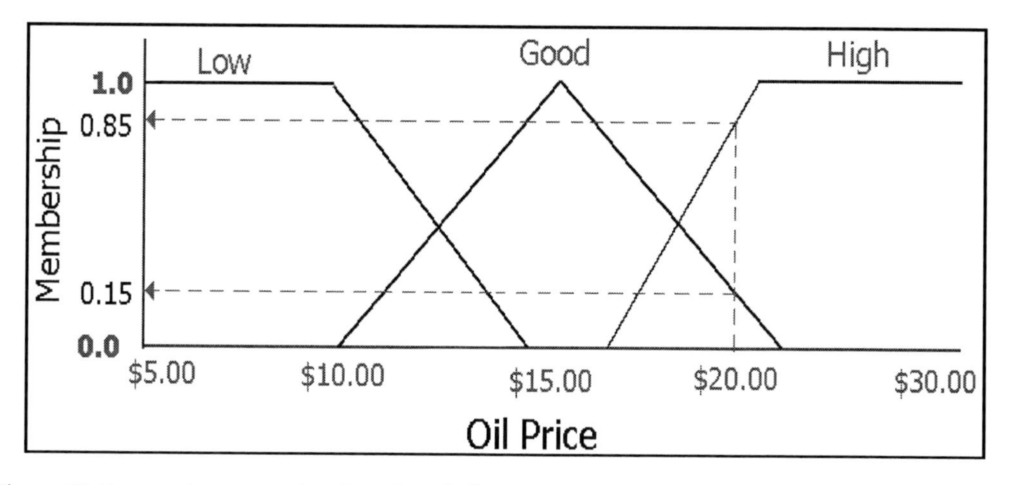

Figure 12. Fuzzy sets representing the price of oil.

The most popular (although not yet standard) form of representing fuzzy set and membership information is as follows:

$$\mu_A(x) = m$$

This representation provides the following information: the membership μ of x in fuzzy set *A* is *m*. According to the Figure 12, when the price of oil is \$20.00 per barrel, it has a membership of 0.15 in the fuzzy set "Good" and a membership of 0.85 in the fuzzy set "High". Using the above notation to represent the oil price membership values,

$$\mu_{Good}(\$20.00) = 0.15 \qquad \mu_{High}(\$20.00) = 0.85$$

3.2. Approximate Reasoning

When decisions are made based on fuzzy linguistic variables (low, good, high) using fuzzy set operators (And, Or), the process is called the approximate reasoning. This process mimics the human expert's reasoning process much more realistically than the conventional expert systems. For example, if the objective is to build a fuzzy expert system to help us make a recommendation on enhanced recovery operations, then we can use the oil price and the company's proven reserves to make such a recommendation. Using the fuzzy sets in Figure 12 for the oil price and the fuzzy sets in Figure 13 for the company's total proven reserves, we try to build a fuzzy system that can help us in making a recommendation on engaging in enhanced recovery operations as shown in Figure 14.

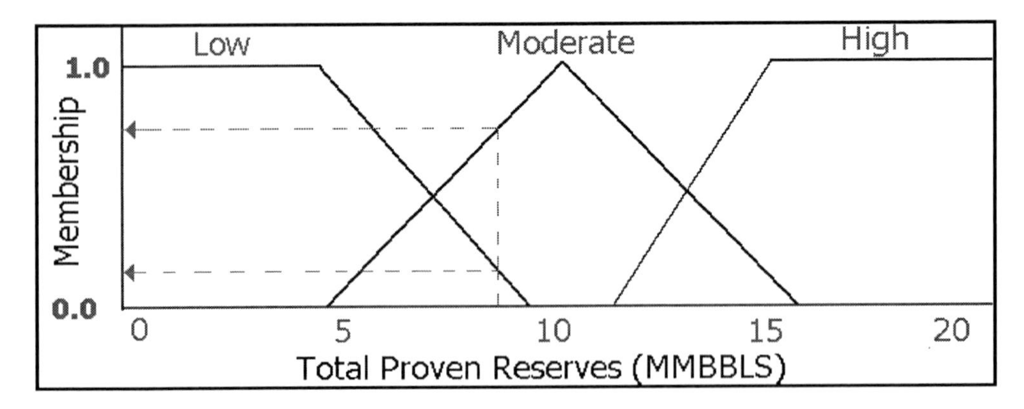

Figure 13. Fuzzy sets representing the total proven reserves.

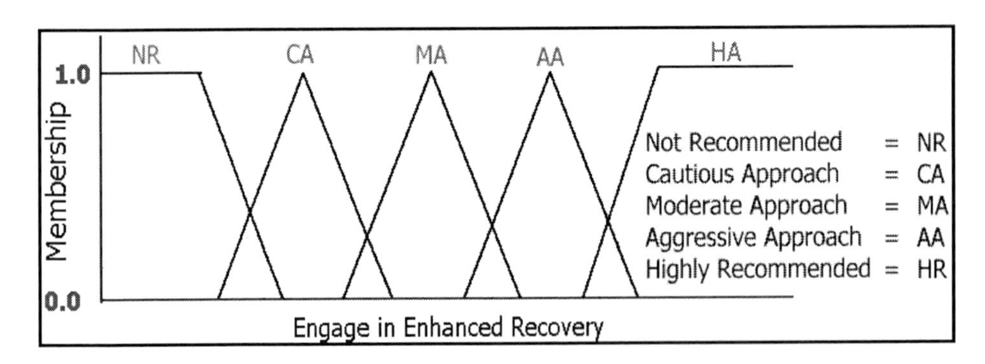

Figure 14. Fuzzy sets representing the decision to engage in enhanced recovery.

The approximate reasoning is implemented through fuzzy rules. A fuzzy rule for the system being explained here can have the following form:

Rule #1: If the Price of Oil is High And the Total Proven Reserves of the company is Low then Engaging in Enhanced Recovery practices is Highly Recommended.

Since this fuzzy system is comprised of two variables and each of the variables consists of three fuzzy sets, the system will include nine fuzzy rules. These rules can be set up in a matrix as shown in Figure 15.

The abbreviations that appear in the matrix above correspond to the fuzzy sets defined in Figure 15. As one can conclude from the above example, the number of rules in a fuzzy system increases dramatically with addition of new variables. Adding one more variable consisting of three fuzzy sets to the above example, increases the number of rules from 9 to 27. This is known as the "curse of dimensionality."

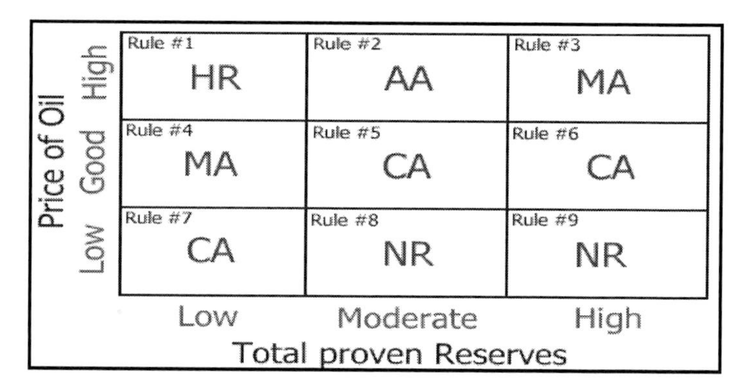

Figure 15. Fuzzy rules for approximate reasoning.

3.3. Fuzzy Inference

A complete fuzzy system includes a fuzzy inference engine. The fuzzy inference helps us build fuzzy relations based on the fuzzy rules that have been defined. During a fuzzy inference process, several fuzzy rules will be fired in parallel.

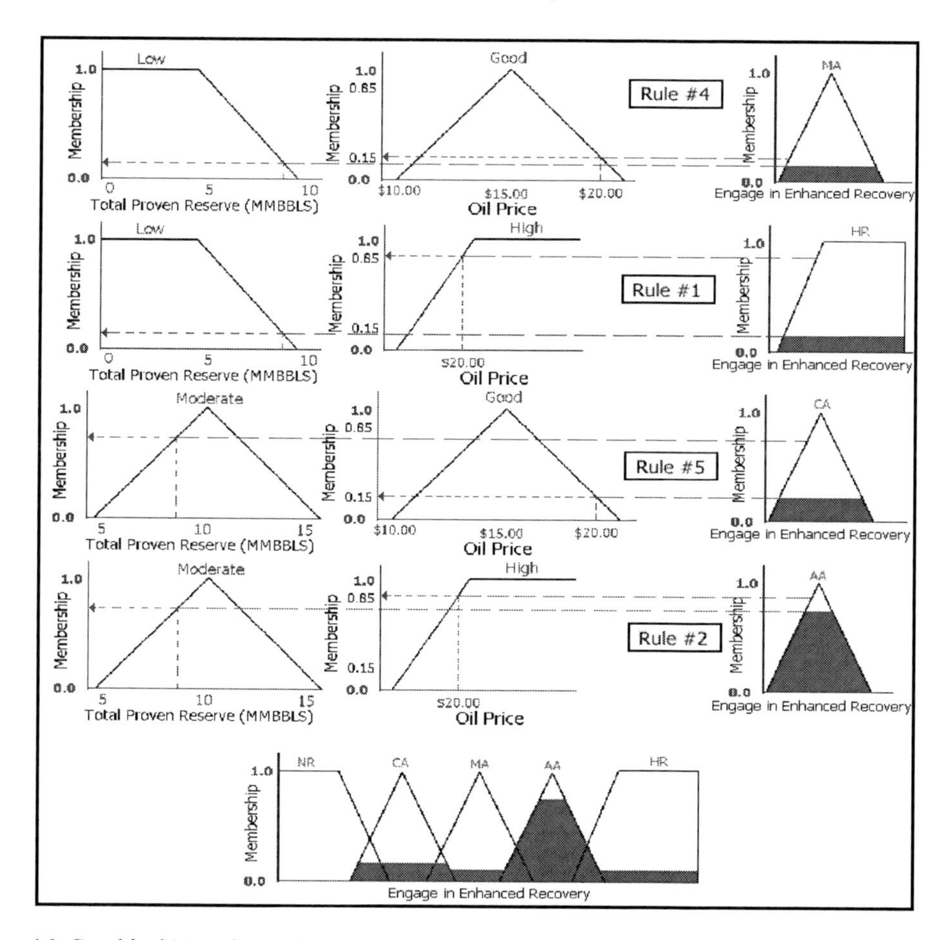

Figure 16. Graphical Mamdany's fuzzy inference.

The parallel rule firing, unlike the sequential evaluation of the rules in the conventional expert system, is much closer to the human reasoning process. Unlike in the sequential process that some information contained in the variables may be overlooked due to the step-wise approach, the parallel firing of the rules allows consideration of all the information content simultaneously. There are many different fuzzy inference methods. We will look at a popular method called the Mamdani's inference method [25] This inference method is demonstrated graphically in Figure 16. In this figure, a case is considered when the price of oil is $20.00 per barrel and the company has approximately 9 MMBBLs of proven reserves. The oil price is represented by its membership in fuzzy sets "Good" and "High", while the total proven reserves is represented in fuzzy sets "Low" and "Moderate". As shown in Figure 16, this causes four rules to be fired simultaneously. According to Figure 15 these are rules #1, #2, #4, and #5. In each rule, the fuzzy set operation "And", the intersection between the two input (antecedents) variables, is evaluated as the minimum and consequently is mapped on the corresponding output (consequent). The result of the inference is the collection of the different fuzzy sets of the output variable as shown on the bottom of the figure.

A crisp value may be extracted from the result as mapped on the output fuzzy sets by defuzzifying the output. One of the most popular defuzzification procedures is to find the center of the mass of the shaded area in the output fuzzy sets.

4. Applications in the Oil and Gas Industry

4.1. Neural Networks Applications

Common sense indicates that if a problem can be solved using conventional methods, one should not use neural networks or any other artificial intelligence technique to solve them. For example, balancing your checkbook using a neural network is not recommended. Although there is academic value to solving simple problems, such as polynomials and differential equations, using neural networks to show its capabilities, they should be used mainly in solving problems that otherwise are very time consuming or simply impossible to solve by conventional methods.

Neural networks have shown great potential for generating accurate analysis and results from large historical databases. The kind of data that engineers may not consider valuable or relevant in conventional modeling and analysis processes. Neural networks should be used in cases where mathematical modeling is not a practical option. This may be due to the fact that all the parameters involved in a particular process are not known and/or the inter-relation of the parameters is too complicated for mathematical modeling of the system. In such cases a neural network can be constructed to observe the system behavior (what types of output is produced as a result of certain set of inputs) and try to mimic its functionality and behavior. In this section few examples of applying artificial neural networks to petroleum engineering related problems is presented.

4.1.1. Reservoir Characterization

Neural networks have been utilized to predict or virtually measure formation characteristics such as porosity, permeability and fluid saturation from conventional well logs

[26, 27, 28] Using well logs as input data coupled with core analysis of the corresponding depth, these reservoir characteristics were successfully predicted for a heterogeneous formation in West Virginia. There have been many attempts to correlate permeability with core porosity and/or well logs using mathematical or statistical functions since the early 1960s [29]. It was shown that a carefully orchestrated neural network analysis is capable of providing more accurate and repeatable results when compared to methods used previously [30].

Figure 17 is a cross-plot of porosity versus permeability for the “Big Injun” formation in West Virginia. It is obvious that there are no apparent correlation between porosity and permeability in this formation. The scatter of this plot is mainly due to the complex and heterogeneous nature of this reservoir.

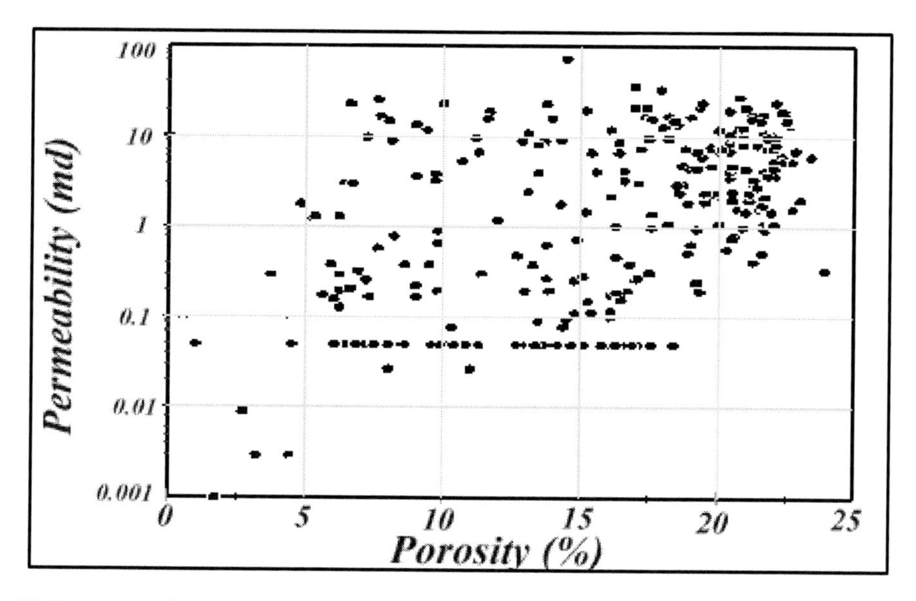

Figure 17. Porosity and permeability cross-plot for Big Injune formation.

Well logs provide a wealth of information about the rock, but they fall short in measurement and calculation of its permeability. Dependencies of rock permeability on parameters that can be measured by well logs have remained one of the fundamental research areas in petroleum engineering. Using the conventional computing tools available, scientists have not been able to prove that a certain functional relationship exists that can explain the relationships in a rigorous and universal manner. Authors suggest that if such dependency or functional relation exists, an artificial neural network is the tool to find it.

Using geophysical well log data as input (bulk density, gamma ray, and induction logs), a neural network was trained to predict formation permeability measured from laboratory core analyses. Log and core permeability data were available from four wells. The network was trained with the data from three wells and attempted to predict the measurements from the fourth well. This practice was repeated twice each time using a different well as the verification well. Figure 18 and Figure 19 show the result of neural network's prediction compared to the actual laboratory measurements. Please note that the well logs and core measurements from these test wells were not used during the training process.

In a similar process well logs were used to predict (virtually measure) effective porosity and fluid saturation in this formation. The results of this study are shown in Figure 20 through Figure 22. In these figures solid lines show the neural network's predictions. The core measurements are shown using two different symbols. The circles are those core measurements that were used during the training process and the triangles are the core measurements that were never seen by the network.

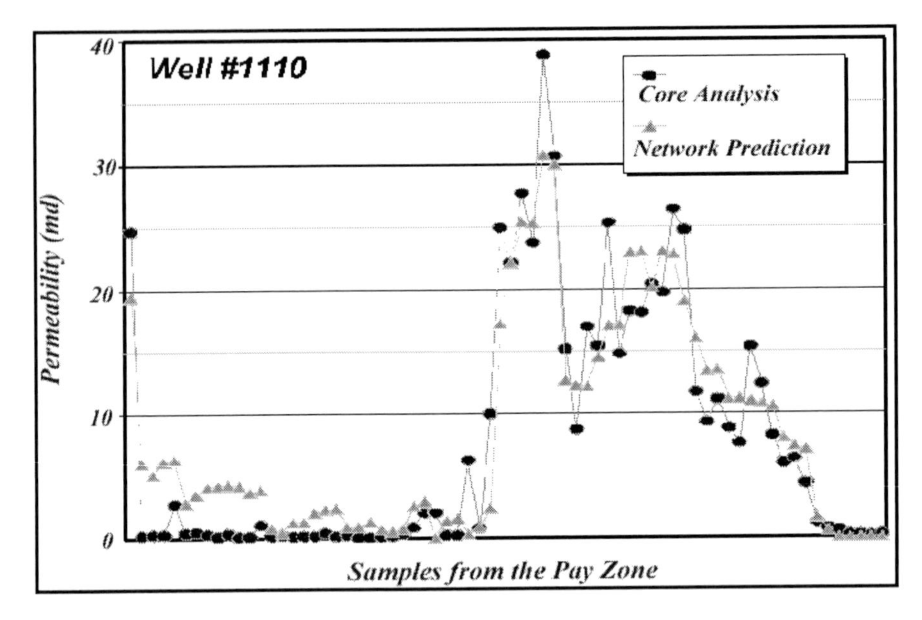

Figure 18. Core and network permeability for well 1110 in Big Injun formation.

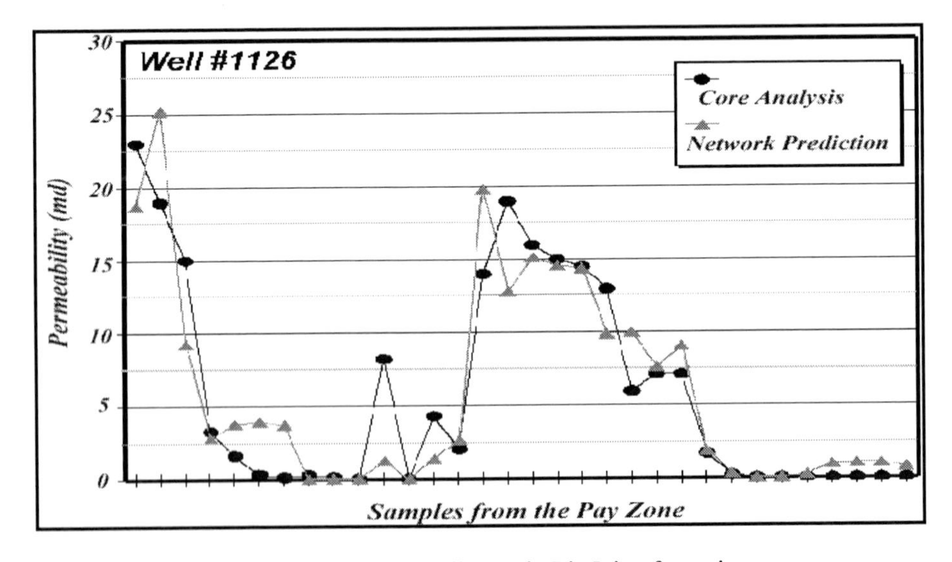

Figure 19. Core and network permeability for well 1126 in Big Injun formation.

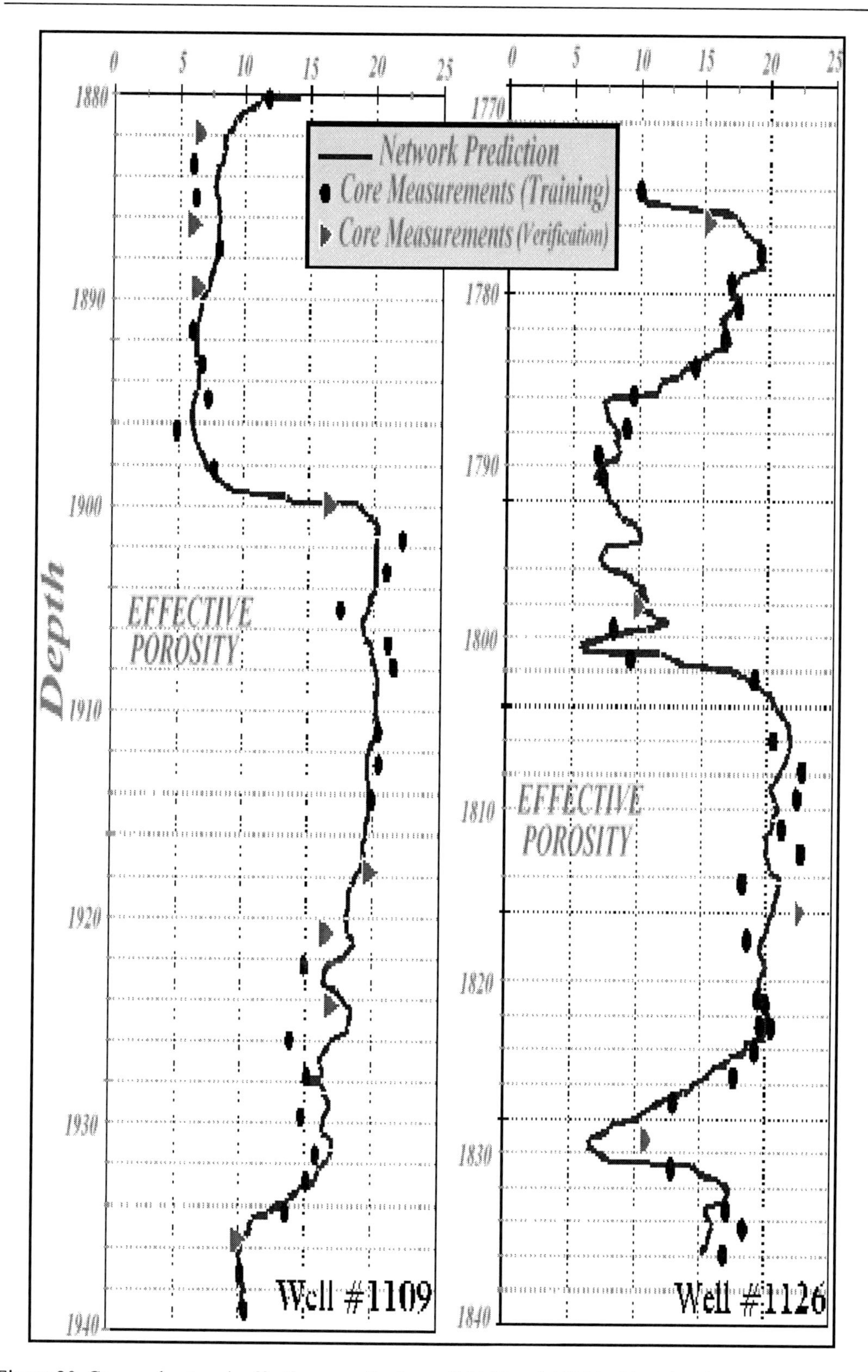

Figure 20. Core and network effective porosity for well 1109 and 1126 in Big Injun formation.

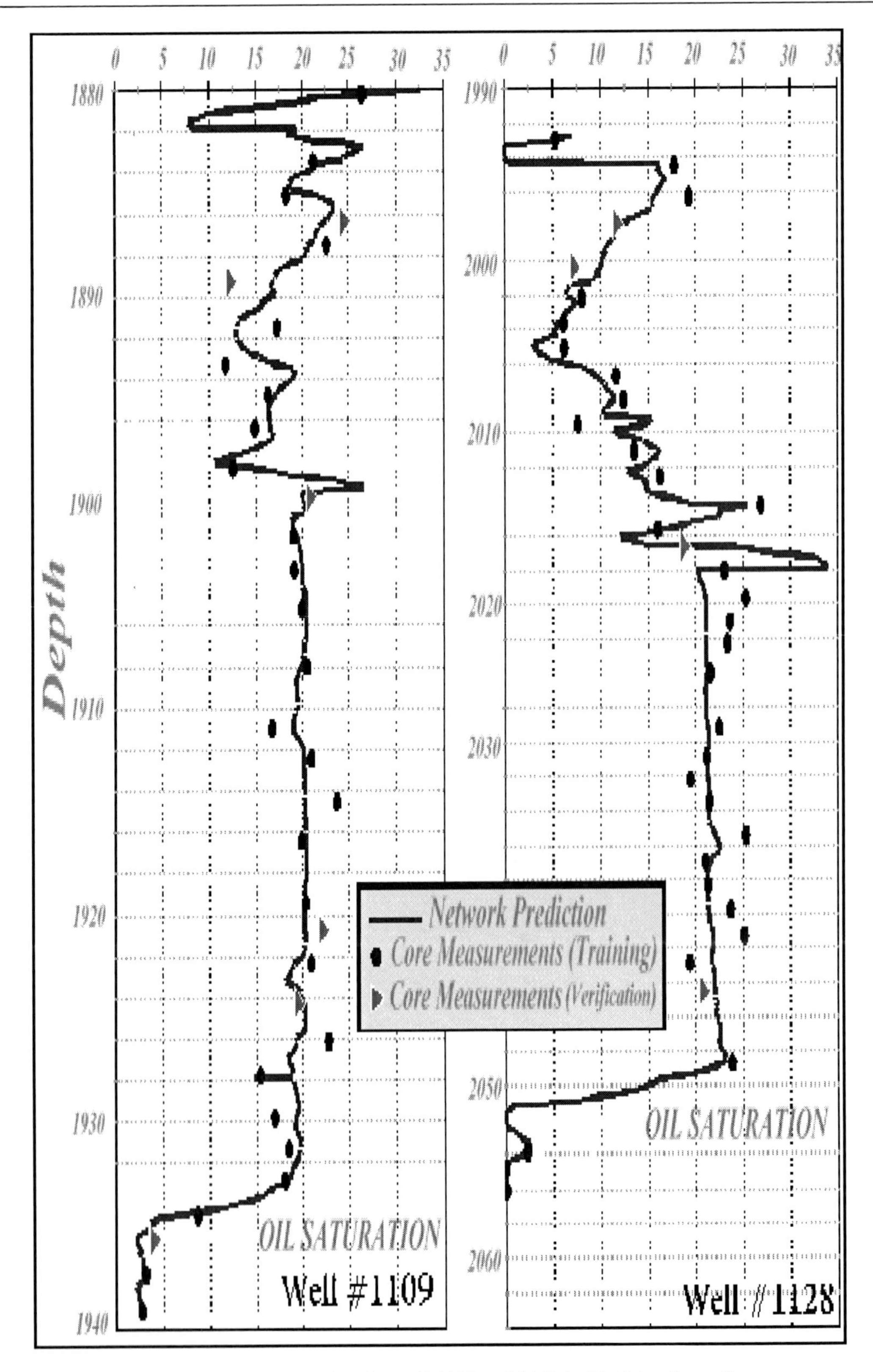

Figure 21. Core and network oil saturation for well 1109 and 1128 in Big Injun formation.

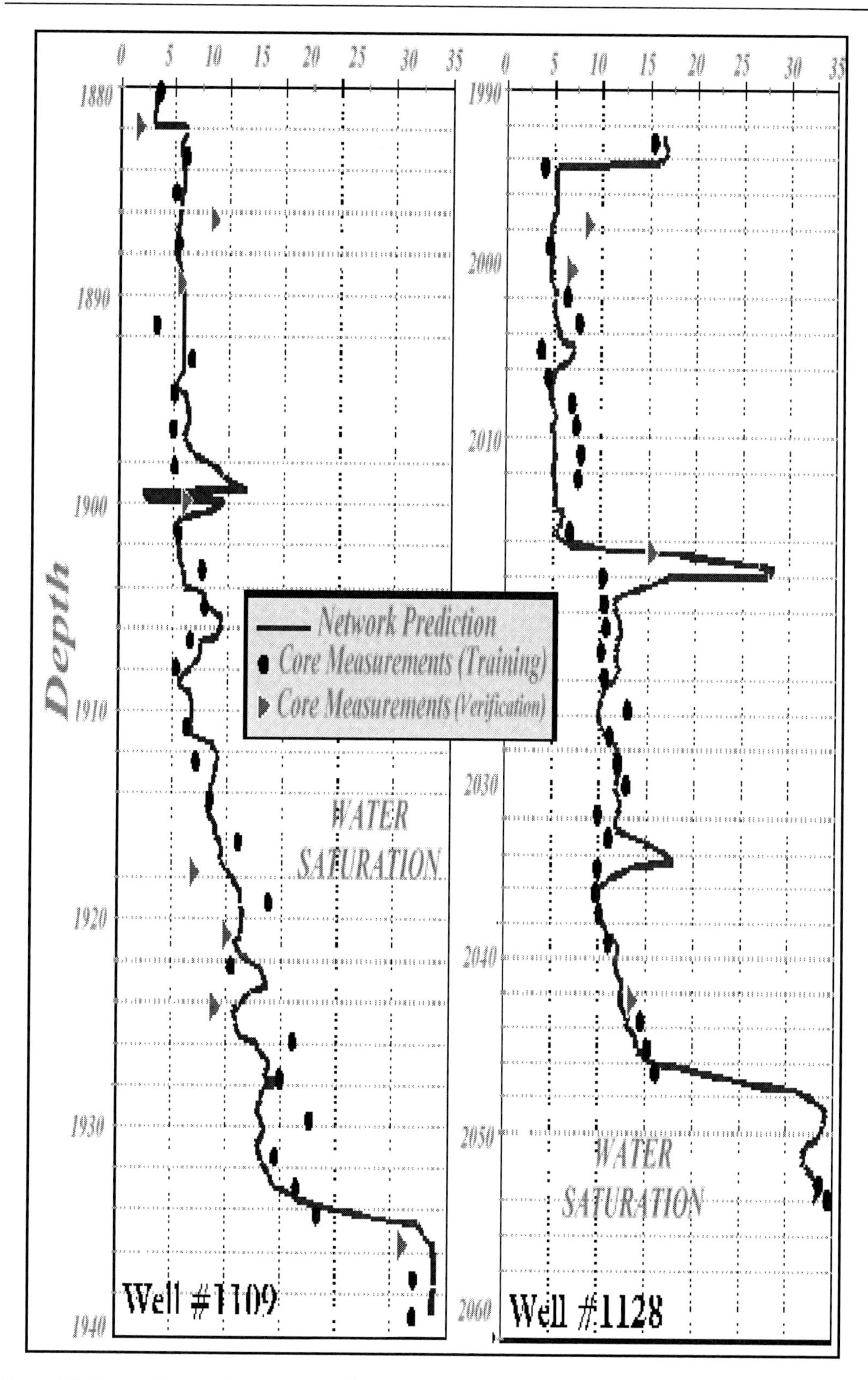

Figure 22. Core and network water saturation for well 1109 and 1128 in Big Injun formation.

4.1.2. Virtual Magnetic Resonance Imaging Logs

Magnetic Resonance Imaging logs are well logs that use nuclear magnetic resonance to measure free fluid, irreducible water (MBVI), and effective porosity (MPHI) accurately. Permeability is then calculated using a mathematical function that incorporates these measured properties. MRI logs can provide information that result in an increase in the recoverable reserve. This takes place simply by including the portions of the pay zone into the recoverable reserve calculations that were excluded during the analysis using only the conventional well logs. MRI logs accomplishes this task by estimating the economically recoverable hydrocarbon (identification of a combination of water and hydrocarbon saturation as well as the reservoir permeability) that has been overlooked. In a recent paper it was shown that neural networks have the potential to be used as an analytical tool for generation of synthetic magnetic resonance imaging logs from conventional geophysical well logs [31]. In this study four wells from different locations in the United States were used to show the potential of this proposed approach. These wells were from Utah, Gulf of Mexico, East Texas and New Mexico. In each case part of the well data is used to train a neural network and the rest of the rest of the well data are used as verification. As it is mentioned in the paper this method is most useful for fields with many wells from which only a handful need to be logged using magnetic resonance imaging tools. These wells can be strategically placed to capture as much reservoir variation as possible. Then a virtual MRI application can be developed based on these wells and applied to the rest of the wells in the field. Figure 23 is an example for such a strategy.

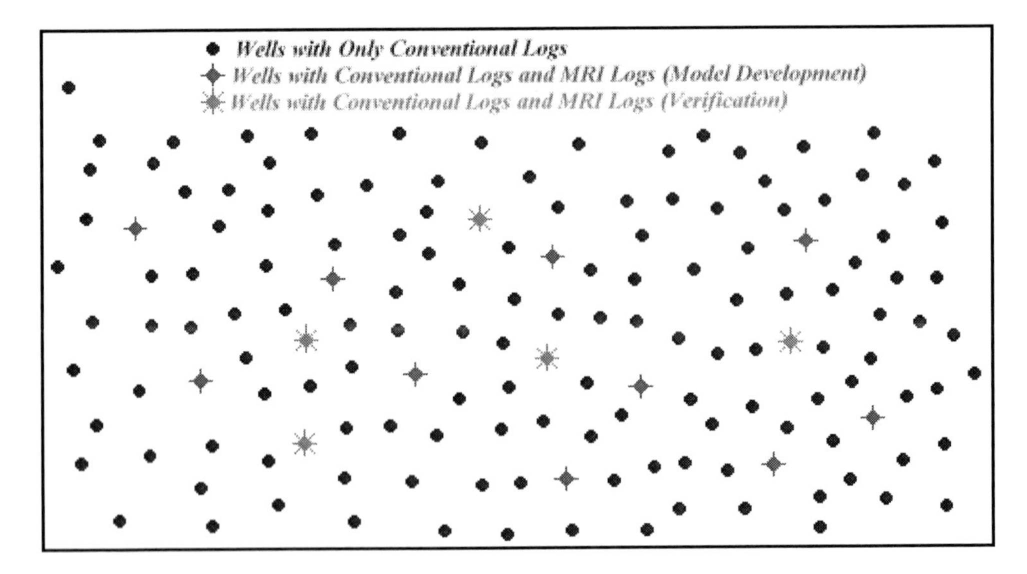

Figure 23. Using virtual MRI log methodology in a typical field.

Table 1 shows the accuracy of this methodology when applied to the four wells being studied. For each well the methodology was applied to three different MRI logs namely MPHI (effective porosity), MBVI (irreducible water saturation), and MPERM (permeability). For each log the table shows the correlation coefficient both for the entire well data set (training data and verification data) and for only the verification data set. The verification data set includes data that had not been seen previously by the network. The correlation coefficient of this methodology ranges from 0.80 to 0.97. As expected, the correlation coefficient for the entire well data set is better than that of the verification data set. This is due to the fact that the training data set is included in the entire well data set and that correlation coefficient for the training data is usually higher than the verification data set.

Table 1. Results of virtual MRI logs for four wells in the United States

Well Location	MRI Log	Data Set	Correlation Coefficient
East Texas	MPHI	Verification Data set	0.941
		Entire Well	0.967
	MBVI	Verification Data set	0.853
		Entire Well	0.894
	MPERM	Verification Data set	0.966
		Entire Well	0.967
Utah	MPHI	Verification Data set	0.800
		Entire Well	0.831
	MBVI	Verification Data set	0.887
		Entire Well	0.914
	MPERM	Verification Data set	0.952
		Entire Well	0.963
Gulf of Mexico	MPHI	Verification Data set	0.853
		Entire Well	0.893
	MBVI	Verification Data set	0.930
		Entire Well	0.940
	MPERM	Verification Data set	0.945
		Entire Well	0.947
New Mexico	MPHI	Verification Data set	0.957
		Entire Well	0.960
	MBVI	Verification Data set	0.884
		Entire Well	0.926

MRI logs are also used to provide a more realistic estimate of recoverable reserve as compared to conventional well logs. Table 2 shows the recoverable reserve calculated using actual and virtual MRI logs. Recoverable reserve calculations based on virtual MRI logs are quite close to those of actual MRI logs since during the reserve calculation a certain degree of averaging takes place that compensates for some of the inaccuracies that are associated with virtual MRI logs. As shown in

Table 2 in all four cases the recoverable reserve calculated using Virtual MRI logs are within 2% of those calculated using actual MRI logs. In the case of the well in the Gulf of Mexico the percent difference is about 0.3%. Although there is not enough evidence to make definitive conclusions at this point, but it seems that recoverable reserve calculated using virtual MRI logs are mostly on the conservative sides.

Table 2. Recoverable reserve calculations using actual and virtual MRI logs

Well Location	MRI Type	Reserve (MMSCF/Acre)	Percent Difference
Texas	Actual	414.58	-1.57
	Virtual	407.95	
New Mexico	Actual	192.73	-1.91
	Virtual	189.05	
Gulf of Mexico	Actual	1,904.93	+0.30
	Virtual	1,910.70	
Utah	Actual	1,364.07	-1.81
	Virtual	1,339.56	

Figure 24 Figure 24 - Virtual MRI log results for the well in East Texas, for verification data set and the entire well data set.and Figure 25 show the comparison between actual and virtual MRI logs for the well located in East Texas.

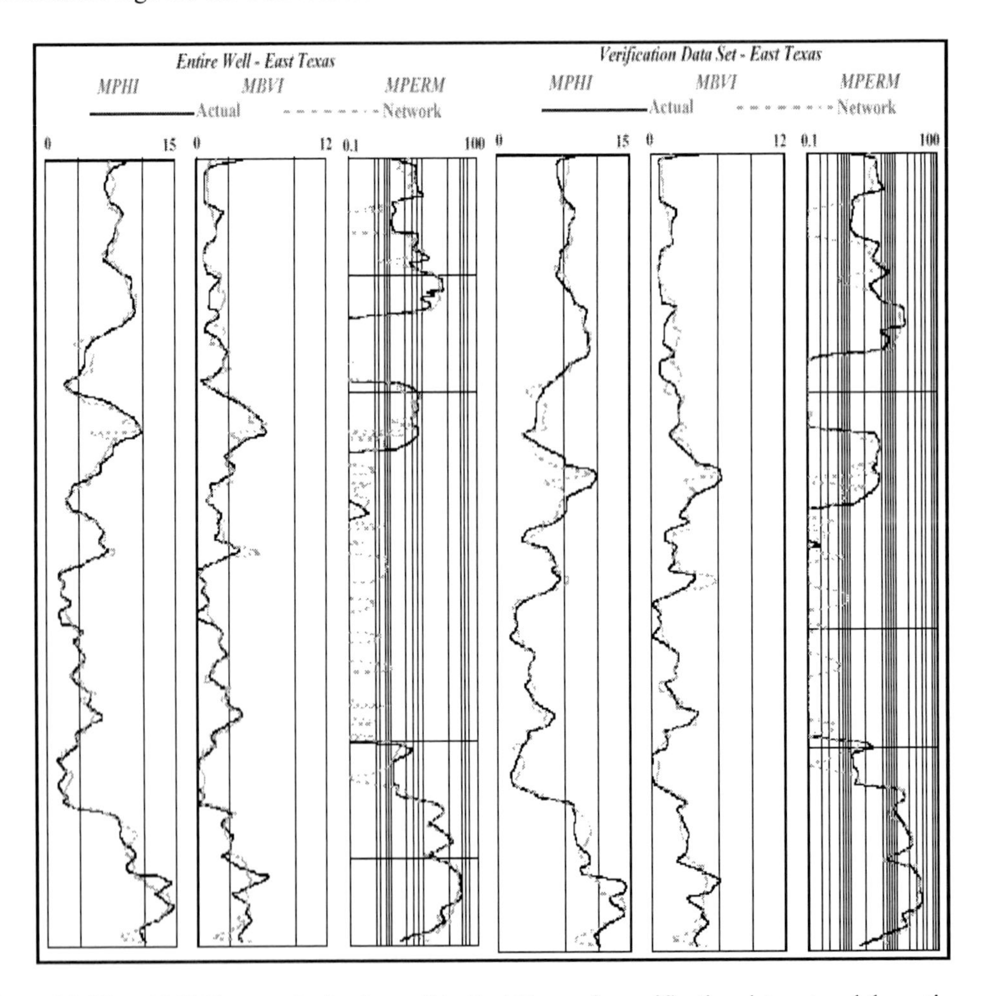

Figure 24. Virtual MRI log results for the well in East Texas, for verification data set and the entire well data set.

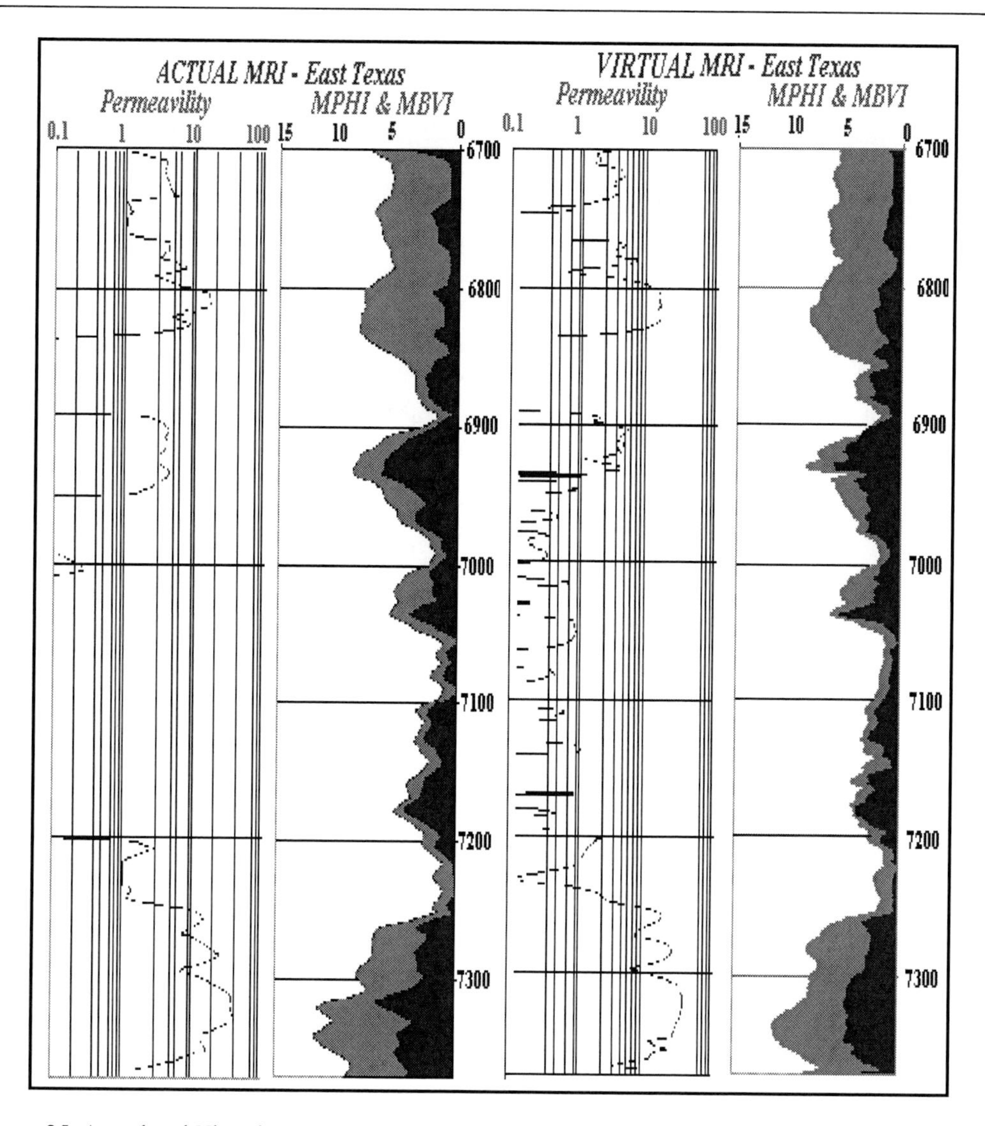

Figure 25. Actual and Virtual MRI log results for the well in East Texas.

There are many more applications of neural networks in the oil and gas industry. They include application to field development [32], two-phase flow in pipes [33, 34], identification of well test interpretation models [35, 36, 37], completion analysis [38, 39], formation damage prediction [40], permeability prediction[41, 42], and fractured reservoirs [43, 44].

4.2. Genetic Algorithms Applications

There have been several applications of genetic algorithms in the petroleum and natural gas industry. The first application in the literature goes back to one of Holland's students named David Goldberg. He applied a genetic algorithm to find the optimum design for gas transmission lines [45]. Since then, genetic algorithms have been used in several other petroleum applications. These include reservoir characterization [46] and modeling [47], distribution of gas-lift injection [48], petrophysics [49] and petroleum geology [50], well test analysis [51], and hydraulic fracturing design [52, 53, 54].

As it was mentioned earlier, virtual intelligence techniques perform best when used to complement each other. The first hybrid neural network/genetic algorithm application in the oil and gas industry was used to design optimum hydraulic fractures in a gas storage field. [53, 54] A brief review of the hybrid neural network/genetic algorithm is presented here.

Virtual intelligence techniques were utilized to design optimum hydraulic fractures for the Clinton Sand in Northeast Ohio. In order to maintain and/or enhance deliverability of gas storage wells in the Clinton Sand, an annual restimulation program has been in place since the late sixties. The program calls for as many as twenty hydraulic fractures and refractures per year. Several wells have been refractured three to four times, while there are wells that have only been fractured once in the past thirty years. Although the formation lacks detailed reservoir engineering data, there is wealth of relevant information that can be found in the well files. Lack of engineering data for hydraulic fracture design and evaluation had, therefore, made use of 2D or 3D hydraulic fracture simulators impractical. As a result, prior designs of hydraulic fractures had been reduced to guesswork. In some cases, the designs were dependent on engineers' intuition about the formation and its potential response to different treatments – knowledge gained only through many years of experience with this particular field. The data set used in this study was collected using well files that included the design of the hydraulic fractures. The following parameters were extracted from the well files for each hydraulic fracture treatment: the year the well was drilled, total number of fractures performed on the well, number of years since the last fracture, fracture fluid, amount of fluid, amount of sand used as proppant, sand concentration, acid volume, nitrogen volume, average pumping rate, and the service company performing the job. The matchup between hydraulic fracture design parameters and the available post fracture deliverability data produces a data set with approximately 560 records.

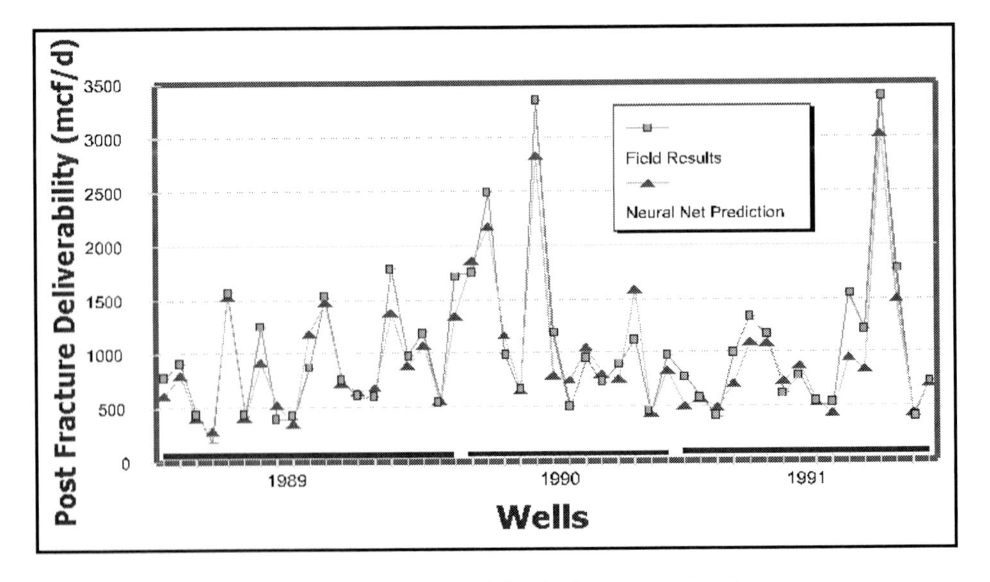

Figure 26. Neural network model's predictive capability in the Clinton Sand.

The first step in this study was to develop a set of neural network models of the hydraulic fracturing process in the Clinton Sand. These models were capable of predicting post fracture deliverability given the input data mentioned above. Figure 26 shows the neural model's predictions compared to actual field results for three years. Data from these years were not used in the training process.

Once the neural network model's accuracy was established, it was used as the fitness function for the genetic algorithm process to form the hybrid intelligent system. The input data to the neural network can be divided into three categories:

- Basic well information
- Well production history
- Hydraulic fracture design parameters such as sand concentration, rate of injection, sand mesh size, fluid type, etc.

From the above categories, only the third (hydraulic fracture design parameters) are among the controllable parameters. In other words, these are the parameters that can be modified for each well to achieve a better hydraulic fracture design. A two-stage process was developed to produce the optimum hydraulic fracture design for each well in the Clinton Sand. The optimum hydraulic fracture design is defined as the design that results in the highest possible post fracture deliverability. Figure 27 is a schematic diagram of the hybrid neuro-genetic procedure.

The neural network for the first stage (neural module #1) is designed and trained to perform a rapid screening of the wells. This network is designed to identify the so-called "dog wells" that would not be enhanced considerably even after a frac job. This way the genetic optimization can be concentrated on the wells that have a realistic chance of deliverability enhancement. The second stage of the process is the genetic optimization routine. This stage is performed on one well at a time. The objective of this stage is to search among all the possible combinations of design parameters and identify the combination of the hydraulic fracture parameters for a specific well that results in the highest incremental post fracture deliverability.

This second stage process (the genetic optimization routine) starts by generating 100 random solutions. Each solution is defined as a combination of hydraulic fracture design parameters. These solutions are then combined with other information available from the well and presented to the fitness function (neural network). The result from this process is the post fracture deliverability for each solution. The solutions are then ranked based on the highest incremental enhancement of the post fracture deliverability. The highest-ranking individuals are identified, and the selection for reproduction of the next generation is made. Genetic operations such as crossover, inversion and mutations are performed, and a new generation of solutions is generated. This process is continued until a convergence criterion is reached.

This process is repeated for all the wells. The wells with highest potential for post fracture deliverability enhancement are selected as the candidate wells. The combination of the design parameters identified for each well is also provided to the operator to be used as the guideline for achieving the well's potential. This process was performed for the wells in Figure 6. The result of the genetic optimization is presented in Figure 28 through Figure 30. Since the same neural networks have generated all the post fracture deliverabilities, it is expected that the post fracture deliverabilities achieved after genetic optimization have the same degree of accuracy as those that were predicted for each well's field result. In these figures, the green bars show the actual PFD of the wells achieved in the field. The red bars show the accuracy of the neural network used as the fitness function in the genetic optimization routine when it is predicting the PDF, given the design parameters used in the field. The blue bars show the PDF resulting from the same neural network that produced the red bars, but with the input design parameters the genetic algorithms proposed.

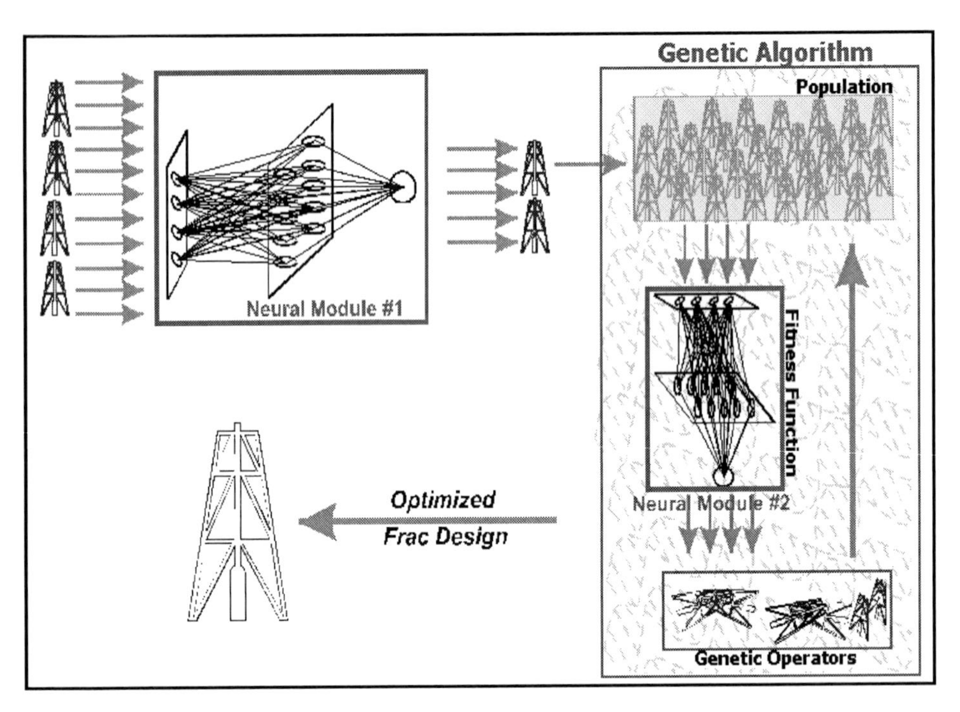

Figure 27. Hybrid neuro-genetic procedure for optimum hydraulic fracture design in the Clinton Sand.

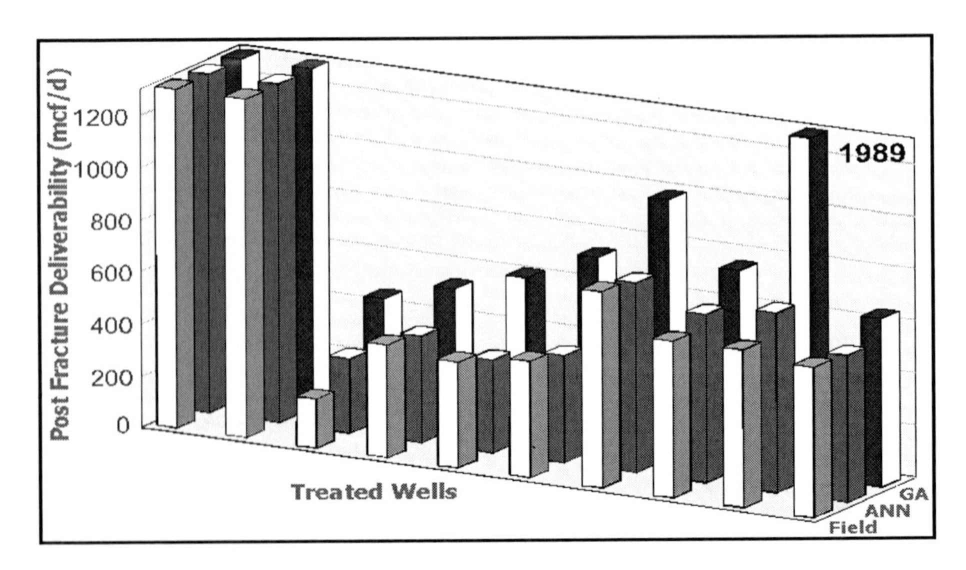

Figure 28. Enhancement in PFD if this methodology had been used in 1989.

Please note that the process indicates that some wells cannot be enhanced, regardless of the modification in the fracture design, while other wells can be enhanced significantly. This finding can have important financial impact on the operation and can help the management make better decisions in allocation of investment.

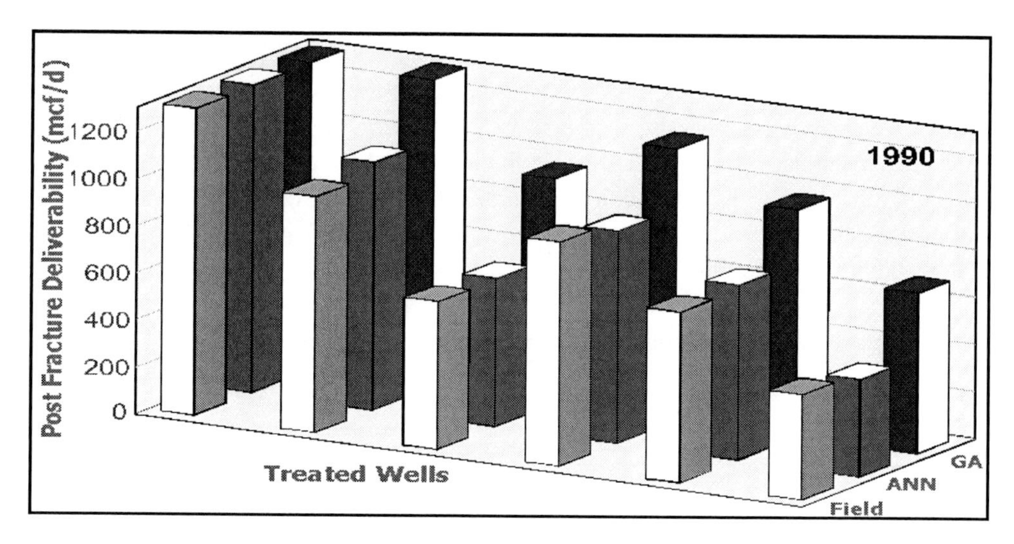

Figure 29. Enhancement in PFD if this methodology had been used in 1990.

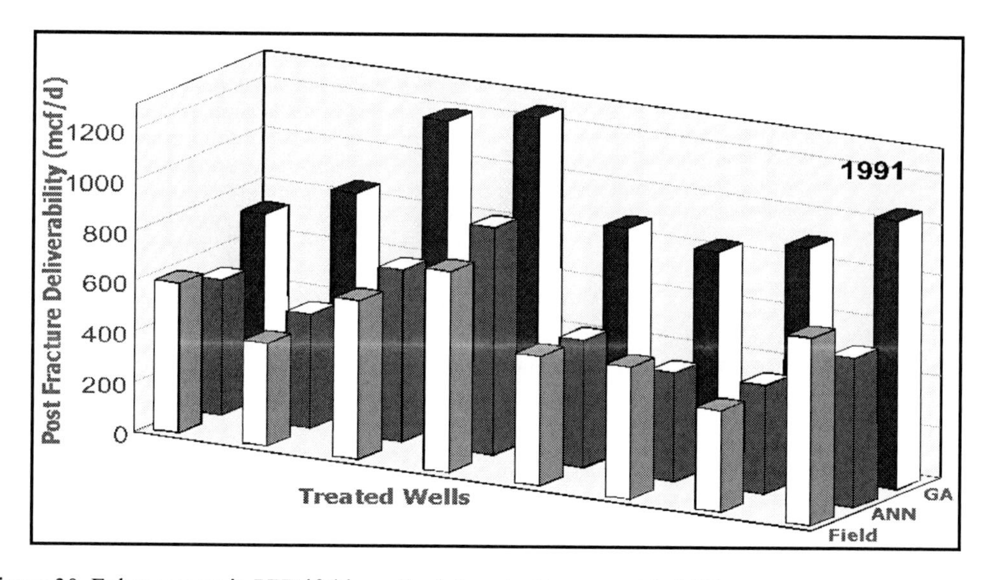

Figure 30. Enhancement in PFD if this methodology had been used in 1991.

In another application, genetic algorithms were used in combination with neural networks to develop an expert hydraulic fracture designer [54]. The intelligent system developed for this purpose is capable of designing hydraulic fractures in detail, providing the pumping schedule in several stages (or in a ramp scheme), and identifying the fluid type and amount, proppant type and concentration, and the pumping rate. It was shown that fracture designs proposed by this intelligent system are comparable to those designed by expert engineers with several years of experience.

4.3. Fuzzy Logic Applications

Fuzzy logic has been used in several petroleum engineering related applications. These applications include petrophysics [55, 56], reservoir characterization [57], enhanced recovery [58, 59] infill drilling [60], decision making analysis [61], and well stimulation [62, 63, 64]. In this section we review an application that incorporates fuzzy logic in a hybrid manner in concert with neural networks and genetic algorithms. In this example of use of the intelligent systems in petroleum engineering, neural networks, genetic algorithms, and fuzzy logic are used to select candidates for restimulation in the Frontier formation in the Green River Basin [64]. As the first step of the methodology, neural networks are used to build a representative model of the well performance in the Frontier formation. Table 3 is a list of input parameters used in the neural network model building process.

Table 3. Input parameters for the neural network analysis

Category	Input Parameter	Comments
Location	X	X coordinates of the well (east-west)
	Y	Y coordinates of the well (north-south)
	KB Elevation	Kelly Bushing Elevation
Reservoir	Permeability	From Type Curve matching analysis
	Drainage Area	From Type Curve matching analysis
	Total Gas-Ft	Sum(Porosity * gas saturation * net pay) (all zones)
Completion	Total H Completed	Total completed thickness (all zones)
	Total No. of Holes	Total number of perforation holes
	Completion Date	Date of well completion
	Number of Zones	Total number of zones completed
Frac	Frac Number	A well may have up to 7 frac jobs
	Fluid type	Gelled oil, ungelled oil, linear gel, cross-linked gel
	Fluid Volume	Total amount of fluid pumped in all fracs
	Proppant Amount	Total amount of proppant pumped in all fracs

Once the training, testing, and validation of the neural networks were completed, the training data set had a correlation coefficient of 0.96, and the verification data set had a correlation coefficient of 0.72. As a by-product of the neural network analysis and by using a methodology called "backward elimination," an attempt was made to identify the most influential parameters in this data set. The results of neural network backward elimination analysis are demonstrated in Figure 31.

In this figure, all four categories of the input data are shown. The most influential category has the lowest R squared. This figure shows that reservoir quality is the most important category, followed by the completion and stimulation categories that seem to be equally important. The location-related input parameters seem to be the least important parameters when compared to others. Note that among all the parameters involved in this analysis only the last three stimulation related parameters (see Table 3) are considered as being controllable.

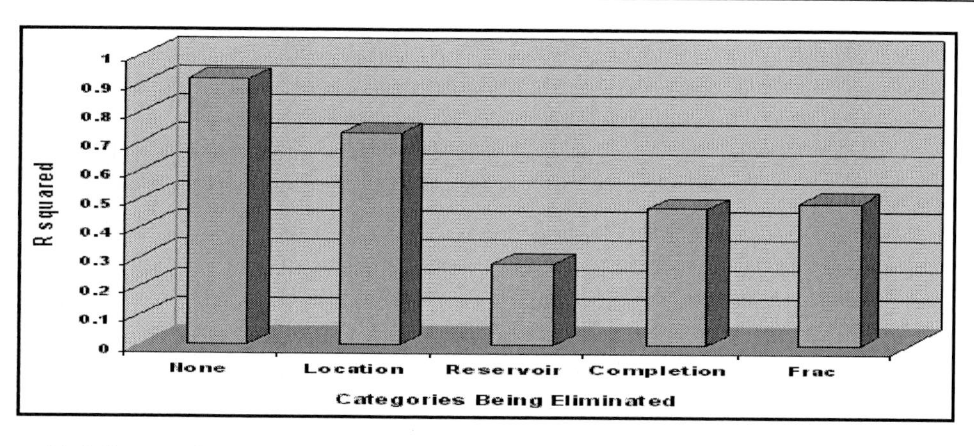

Figure 31. Influence of parameters in the stimulation process in Frontier formation.

This brings us to the second step of the analysis that involves the genetic optimization of the stimulation parameters. In this step, the last three input parameters shown in Table 3 (namely fluid type, total fluid volume, and total proppant amount) are used in the optimization process. Using the neural network model developed in the first step of the analysis as the "fitness" function of the evolution process, the algorithm searches through all possible combinations of the aforementioned three stimulation parameters and tries to find the combination that results in the highest five-year cumulative production (5YCum). This process is repeated for every well individually. The difference between the optimized 5YCum and the actual 5YCum is considered to be the potentially missed production that may be recovered by restimulation. The outcome of this process is called the potential 5YCum and is used as one of the three inputs into step three which is the fuzzy decision support system using approximate reasoning.

Step three is a three-input, one-output, fuzzy system. The inputs include the above-mentioned potential 5YCum, a calculated parameter called Fracs per Zone (FPZ), and pressure.

Fracs/Zone	Pressure Low: Potential Five Year Cum. Low	Med.	High	Pressure Medium: Potential Five Year Cum. Low	Med.	High	Pressure High: Potential Five Year Cum. Low	Med.	High
Low	NO T	Maybe FT	Maybe FT	Maybe T	Yes FT	Yes FT	Yes T	Yes VT	Yes VT
Med.	NO VT	NO T	Maybe FT	NO FT	Maybe T	Yes FT	Maybe VT	Yes T	Yes VT
High	NO VT	NO VT	NO T	Maybe FT	NO FT	Maybe T	Maybe VT	Maybe VT	Yes T

Figure 32. Rules used in the fuzzy decision support system.

The engineers in the field brought this parameter to our attention. They mentioned that there are wells that have been completed in all zones (there can be as many as 7 zones present) but only one hydraulic fracture has been performed. In other words, the ratio of the number of treatments performed to the total number zones completed is an important factor. We also found that long-term pressure surveys had been performed in 1995 on many wells. The issue with the pressure surveys is that the shut-in time and the depth where the pressure readings were taken were not consistent throughout the field. This introduces serious imprecision in the pressure values as a comparative value from well to well. Therefore, all the three input parameters were subjected to fuzzy sets using low, moderate, and high fuzzy sets. The output of the fuzzy system is the degree of which a well is a candidate for restimulation. The output fuzzy sets include: 1) the well is a candidate, 2) the well may be a candidate, and 3) the well is not a candidate. The system includes 27 fuzzy rules that are qualified using a set of three truth functions. Figure 32 shows the 27 rules with truth qualification for the fuzzy systems. Figure 33 shows the truth qualification functions used for the approximate reasoning implementation in the fuzzy system. As demonstrated in this figure, each rule can be true, fairly true, or very true.

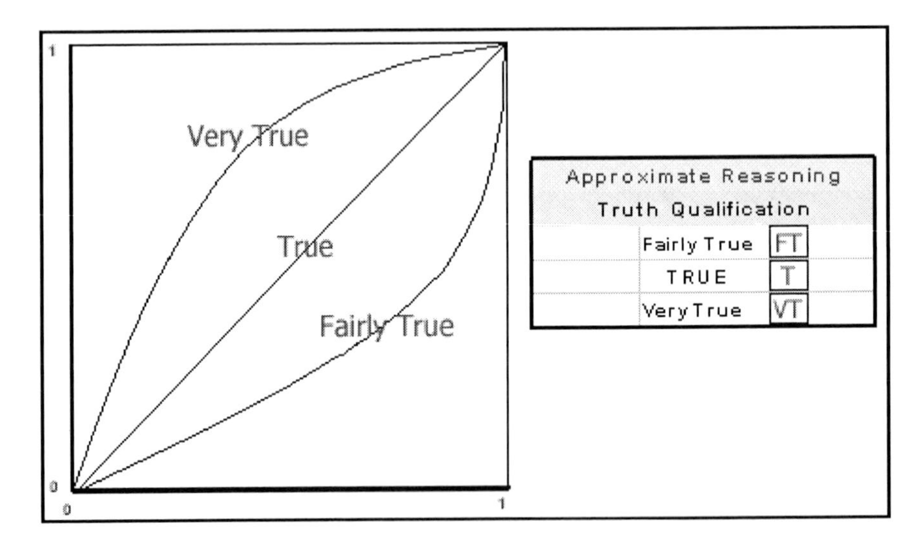

Figure 33. Truth qualification for the fuzzy rules.

Using this three-step process, all the wells (belonging to a particular operator) in the Frontier formation was processed. A list of restimulation candidates was identified.

4.3.1. Results

It should be noted that the intelligent systems approach for this application was modified as a result of its application to three different formations, two in the Rocky Mountains and one in East Texas. The fuzzy decision support system was the most recent addition to the process. The new and improved intelligent systems approach, that included the fuzzy logic component, picked well GRB 45-12 as candidate #20, while this well was missed as a candidate prior to the addition of fuzzy logic to this procedure. An engineer with several years of experience in this field also had suggested this well as a candidate. The fuzzy decision support system was able to capture the engineer's knowledge and use it in an automatic process for all the wells in the study. Figure 34 shows the result of restimulation on Well GRB 45-12.

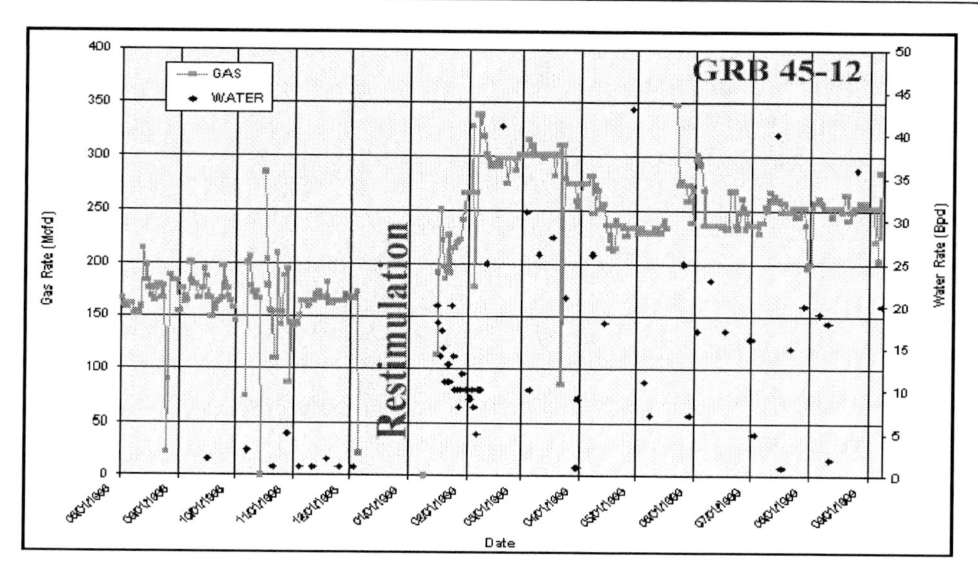

Figure 34. Gas and water production for well GRB-45-12 before and after restimulation.

REFERENCES

[1] Zaruda, J. M., Marks, R. J. and Robinson, C. J. *Computational Intelligencem Imitating Life,*. Poscataway, NJ : IEEE Press, 1994.

[2] Eberhart, R., Simpson, P. and Dobbins, R. *Computational Intelligence PC Tools.* Orlando, FL : Academic Press, 1996.

[3] McCulloch, W. S. and Pitts, W. *A Logical Calculus of Ideas Immanent in Nercous Activity.* 1943. pp. 115-133. Vol. 5.

[4] Rosenblatt, F. The Perceptron: *Probabilistic Model for Information Storage and Organization in the Brain.* 1958. pp. 386-408. Vol. 65.

[5] Widrow, B. Generalization and Information Storage in Networks if Adeline Neurons. [book auth.] M. C. Yovits, G. T. Jacobi and G. D. Goldstein. *Self-Organizing Systems.* Chicago : s.n., 1962, pp. 435-461.

[6] Minsky, M. L. and Papert, S. A. P. *Perceptrons.* Cambridge, MA : MIT Press, 1969.

[7] Hertz, J., Krogh, A. and Palmer, R.G. *Introduction to the Theory of Neural Computation.* Redwood City, CA : Addison-Wesley Publishing Company, 1991.

[8] Rumelhart, D. E. and McCelland, J. L. *Parallel Distributed Processing, Exploration in the microstructure of Cognition,*. Cambridge, MA : MIT Press, 1986. Vol. 1: Foundations.

[9] Stubbs, D. *Neurocomputers.* 1988. Vol. 5.

[10] Fausett, L. Fundamentals of Neural Networks, Archintectures, algorithms, and applications. *Englewood Cliffs, NJ* : Prentice Hall, 1994.

[11] Barlow, H. B. *Unsupervised Learning.* 1988. pp. 295-311. Vol. 1.

[12] McCord-Nelson, M. and Illingworth, W. T. *A Practical Guide to Neural Nets.* Reading, MA : Addison-Wesley Publishing, 1990.

[13] Mayr, E. Toward a new Philosophy of Biology: *Observations of an Evolutionist.* Cambridge, MA : Belknap Press, 1988.

[14] Koza, J. R. *Genetic Programming, On the Programing of Computers by Means of Natural Selection.* Cambridge, MA : MIT Press, 1992.

[15] Fogel, D. B. Evolutionary Computation, toward a New Philosophy of Machine Intelligence. *Picataway, NJ* : IEEE Press, 1995.

[16] Michalewicz, Z. *Genetic Algorithms + Data Structure = Evolution Programs*. New York, NY : Springer-verlag, 1992.

[17] Freeman, E. The Relevance of Charles Pierce. La Shall, IL : *Moinst Library of Philosophy*, 1983. pp. 157-158.

[18] Lukasiewicz, J. *Elements of Mathematical Logic*. New York, NY : The MacMillan Company, 1963.

[19] Black, M. Vagueness: *An Exercise in Logicak Analysis. Philosophy of Science*. 1937, Vol. 4, pp. 437-455.

[20] Zadeh, L. A. *Fuzzy Sets. Information and Control*. 1965, Vol. 8, pp. 338-353.

[21] Kosko, B. Fuzzy Thinking. New York, NY : *Hyperion*, 1991.

[22] Kosko. Neural Networks and Fuzzy Systems. *Englewood Cliffs, NJ* : Prentice Hall, 1992.

[23] McNeill, D. and Freiberger, P. *Fuzzy Logic*. New York, NY : Simon and Schuster, 1993.

[24] Ross, T. *Fuzzy Logic with Engineering Applications*. New York, NY : McGraw-Hill Inc, 1995.

[25] Jamshidi, M. and et.al. Fuzzy Logic and Controls: Software and Hardware Applications. *Englewood Cliffs, NJ* : Prentice Hall, 1993.

[26] Mohaghegh, S., et al. Design and Development of an Artificial Neural Network for Estimation of Formation Permeability. *SPE Computer Applications*. December 1995, pp. 151-154.

[27] Mohaghegh, S., Arefi, R. and Ameri, S. Petroleum Reservoir Characterization with the Aid of Artificial Neural Networks. *Journal of Petroleum Science and Engineeing*. 1996, Vol. 16, pp. 263-274.

[28] Mohaghegh, S., Ameri, S. and Arefi, R. Virtual Measurement of Heterogeneous Formation Permeability Using Geophysical Well Log Responses. *The Log Analysi.* March/April 1996, pp. 32-39.

[29] Balan, B., Mohaghegh, S. and Ameri, S. State-of-the-art in Permeability Determination from Well Log Data, Part 1: A Comparative Study, Model Development. *SPE Eastern Regional Conference*, SPE 30978. September 1995.

[30] Mohaghegh, S., Balan, B. and Ameri, S. State-of-the-art in Permeability Determination from Well Log Data: Part 2: Verifiable, Accurate Permeability Prediction, the Touch-Stone of All Models. *SPE Eastern Regional Conference*, SPE 30979. September 17-21, 1995.

[31] Mohaghegh, S., Richardson, M. and Ameri, M. Virtual Magnetic Resonance Imaging Logs: Genration of Synthetic MRI Logs From Conventional Well Logs. *SPE Eastern Regional Conference,* SPE 51075. November 9-11, 1998.

[32] Doraisamy, H., Ertekin, T. and Grader, A. Key Parameters Controlling the Performance of Neuro Simulation Applications in Field Development. SPE Eastern Regional Conference, SPE 51079. November 9-11, 1998.

[33] Ternyik, J., Bilgesu, I. and Mohaghegh, S. Virtual Measurement in Pipes, Part 2: Liquid Holdup and Flow Pattern Correlation. *SPE Eastern Regional Conference and Exhibition*, SPE 30976. September 19-21, 1995.

[34] Ternyik, J., et al. Virtual Measurement in Pipes, Part 1: Flowing Bottomhole Pressure Under Multi-phase Flow and Inclined Wellbore Conditions. *SPE Eastern Regional Conference and Exhibition*, SPE 30975. September 19-21, 1995.

[35] Sung, W., Hanyang, U. and Yoo, I. Development of HT-BP Neural Network System for the Identification of Well Test Interpretation Model. *SPE Eastern Regional Conference and Exhibition*, SPE 30974. September 19-21, 1995.

[36] Al-Kaabi, A. and Lee, W. J. Using Artificial Neural Nets to Identify the Well Test Interpretation Model. *SPE Formation Evaluation*. September 1993, pp. 233-240.

[37] Juniradi, I. J. and Ershaghi, I. Complexities of Using Neural Networks In Well Test Analysis of Faulted Reservoir. *SPE Western Regional Conference and Exhibition*. March 26-28, 1993.

[38] Shelly, R., et al. Granite Wash Completion Optimazation with the Aid of Artificial Neural Networks. *Gas Technology Symposium*, SPE 39814. March 15-18, 1998.

[39] Shelly, R., et al. Red Fork Analysis with the Aid of Artificial Neural Networks. *Rocky Mountain Regional Meeting / Low Permeabiloity Reservoir Symposium*. April 5-8, 1998.

[40] Nikravesh, M., et al. Prediction of Formation Damage During the Fluid Injection into Fractured Low Permeability Reservoirs via Neural Networks. *Formation Damage Symposium,* SPE 31103. February 16-18, 1996.

[41] Wong, P. M., Henderson, D. J. and Brooks, L. J. Permeability Determination using Neural Networks in the Ravva Field, Offshore India. *SPE Reservoir Evaluation and Engineering*. 1998, Vol. 2, pp. 99-104.

[42] Wong, P. M., Taggart, I. J. and Jian, F. X. A Critical Comparison of Neural Networks and Discriminant Analysis in Lithofacies, Porosity and Permeability Predictions. *Journal of Petroleum Geology*. 1995, Vol. 2, pp. 191-206.

[43] Ouense, A., et al. Use of Neural Networks in Tight Gas Fractured Reservoirs: Application to San Juan Basin. *Rocky Mountain Regional Meeting / Low Permeability Reservoir Symposium*. April 1998, pp. 5-8.

[44] Zellou, A., Ouense, A. and Banik, A. Improved Naturally Fractured Reservoir Characterization Using Neural Networks, Geomechanics and 3-D Seismic. *SPE Annual Technical Conference and Exhibition*. October 22-25, 1995.

[45] Goldberg, D. E. Computer Aided Gas Pipeline Operartion Using Genetic Algorithms and Rule Learning,. University of Michigan, Ann Arbor, MI : PhD Dissertation, 1983.

[46] Guerreiro, J.N.C and et.al. Identification of Reservoir Heterogeneties Using Tracer Breakthrough Profiles and Genetic Algorithms. *Latin American and Caribean Petroleum Engineering Conference and Exhibition*, SPE 39066. August 1997.

[47] Sen, M.K. and et.al. Stochastic Reservoir Modeling Using Simulated Annealing and Genetic Algorithm. *SPE Annual Technical Conference and Exhibition*, SPE 24754. October 4-7, 1992.

[48] Martinez, E. R. and et.al. Application of Genetic Algorithm on *the Distribution of Gas-Lift Injection. SPE* Annual Technical Conference and Exhibition. September 25-28, 1994.

[49] Fang, J. H. and et.al. Genetic Algorithm and Its Application to Petrophisics. *Unsolicited, SPE* 26208. 1992.
[50] Hu, L. Y. and et.al. Random Genetic Simulation of the Internal Geometry of Deltaic Sandstone Bodies. *SPE Annual Technical Conference and Exhibition* , SPE 24714. October 4-7, 1992.
[51] Yin, et al. An Optimum Method of Early-Time Well Test Analysis - Genetic Algorithm. *International Oil and Gas Conference and Exhibition*, SPE 50905. November 2-6, 1998.
[52] Mohaghegh, S., et al. A Hybrid, Neuro-Genetic Approach to Hydraulic Fracture Treatment Design and Optimization. *SPE Annual Technical Conference and Exhibition*, SPE 36602. October 6-9, 1996.
[53] Mohaghegh, S., Platon, V. and Ameri, S. Candidate Selection for Stimulation of Gas Storage Wells Using Available Data with Neural Networks and Genetic Algorithms. *SPE Eastern Regional Meeting*, SPE 51080. November 9-11, 1998.
[54] Mohaghegh, S., Popa, A. S. and Ameri, S. Intelligent Systems Can Design Optimum Fracturing Jobs. *SPE Eastern Regional Conferene and Exhibition*, SPE 57433. October 21-22, 1999.
[55] Zhanggui and et.al. Integration of Fuzzy Methods into Geostatistics for Petrophysical Property Distribution. *SPE Asia Pacific Oil and Gas Conference and Exhibition.* October 12-14, 1998.
[56] Chen, H. C., et al. Novel Approaches to the Determination of Archie Parameters II: Fuzzy Regression Analysis. *Unsolicited, SPE* 26288. 1993.
[57] Zhou, et al. Determining Reservoir Properties in Reservoir Studies Using a Fuzzy Neural Network. *SPE Annual Technical Conference and Exhibition*. October 3-6, 1993.
[58] Chung, T., Carrol, H. B. and Lindsey, R. Application of Fuzzy Expert Systems for EOR Project Risk Analysis. *SPE Annual Technical Conference and Exhibition*, SPE 30741. October 22-25, 1995.
[59] Nikravesh, M., et al. Field-wise Waterflood Management in Low Permeability, Fractured Oil Reservoirs: Neuro-Fuzzy Approach. *SPE International Thermal Operations and Heavy Oil Symposium.* February 10-12, 1997.
[60] Wu, C. H., Lu, G. F. and Yen, J. Statistical and Fuzzy Infill Drilling Recovery Models for Carbonate Reservoirs. *Middle East Oil Conference and Exhibition*. March 17-22, 1997.
[61] Yong, et al. Fuzzy-Grey-Element Relational Decision- Making Analysis and Its Application. *SPE India Oil and Gas Conference and Exhibition*, SPE 39579. February 17-19, 1998.
[62] Xiong and Hongjie. An Investigation into the Application of Fuzzy Logic to Well Stimulation Treatment Design. *SPE Permian Basin Oil and Gas Recovery Conference*. March 16-18, 1994.
[63] Rivera, V. P. Fuzzy Logic Controls Pressure in Fracturing Fluid Characterization Facility. *SPE Petroleum Computer Conference*, SPE 28239. 1994.
[64] Mohaghegh, S., Reeves, S. and Hill, D. Development of an Intelligent Systems Approach to Restimulation Candidate Selection. *SPE Gas Technology Symposium*, SPE 59767. April 2000.

In: Artificial Intelligence
Editor: Brent M. Gordon, pp. 39-70

ISBN 978-1-61324-019-9

Chapter 2

An Artificial Intelligence Approach for Modeling and Optimization of the Effect of Laser Marking Parameters on Gloss of the Laser Marked Gold

V. R. Adineh and A. Alidoosti*
Department of Mechanical Engineering, Saveh Branch,
Islamic Azad University, Saveh, Iran

Abstract

One of the most recent applications of laser marking process is in the manufacturing of decorative gold. Gloss of the final gold marked is a criterion to evaluate the quality of product in terms of aesthetics appearance. This property essentially affected by various laser marking parameters such as laser power, feed rate (speed), Q-switch frequency (QSF) and pulse width (PW). In this paper, an adaptive neuro-fuzzy inference system (ANFIS) technique and artificial neural networks (ANNs) were utilized to model the effect of the mentioned parameters on the gloss of the laser marked gold. Both models were trained with experimental data. The results of this part of study indicated that ANNs had better outcomes compared to ANFIS. The best model was a cascade-forward backpropagation(CFBP) network, with various threshold functions (TANSIG-TANSIG-LOGSIG) and 9/8 neurons in the first/second hidden layers. Afterwards, in order to find the mentioned parameters of laser marking process,which maximize the gloss of the gold, the genetic algorithm (GA) and particle swarm optimization (PSO)were utilized and the best model was presented to the GA and PSO as the objective function. After the optimization, results of this part revealed that GA had better outcome compared to PSO so thatthe calculated gloss effect increases by 15% and the measured value increases by 12% in an experiment as compared to a non-optimized case.

---

* E-mail:v.r.adineh@iau-saveh.ac.ir.

Keywords: Gloss of the Laser marked Gold, Artificial Neural Networks, Adaptive Neuro-Fuzzy Inference Systems, Genetic Algorithm, Particle Swarm Optimization.

1. INTRODUCTION

Using a laser to mark or code information on a product, laser marking, is one of the most industrial applications of lasers [1-2]. Laser marking provides an elegant solution when a clean, fast, non-contact marking process is required to produce an indelible high-quality mark [3-4].

Laser marking is a surface process which includes laser coding (such as alphanumeric code imprinting on the surface of product to describe date of manufacture and etc), functional marking (such as scribing gradation lines on a syringe) or decorative marking (such as engraving a graphic image on a surface) [1-2]. One of the recent applications of the decorative marking is in the jewelry industrial, which its goal is to decorate the graphical images on the gold.

Gloss is an attribute of visual appearance that originates from the geometrical distribution of light reflected by the surface [5]. For many products, a homogenous and consistent gloss isconsidered as a decorative quality parameter and is influenced by many production/process parameters [6]. Inthe other words, gloss is often used as a criterion to evaluate the quality of a product, especially in the case of products where the aesthetic appearance is of importance[7]. As mentioned before, in the laser marking, this property affected by laser power, feed rate (speed), Q-switch frequency (QSF) and pulse width (PW).

Modeling of systems is of fundamental importance in almost all fields. This is because models enable us to understand a system better, simulate and predict the system behavior and hence help us in the optimization of system parameters. Widely developed linear models are applied in the different areas of engineering, but most of the real time systems are ill defined and uncertain in nature. As a result, system modeling based on the traditional linear systems is not appropriate for complicated systems. Non-traditional systems, namely soft computing is a collection of methodologies like fuzzy systems (FS), artificial neural networks (ANNs), genetic algorithm (GA), particle swarm optimization (PSO) and so forth, was designed to tackle imprecision and uncertainty involved in a complex nonlinear system. The evolution of soft computing techniques has helped in understanding the various aspects of nonlinear systems and thereby making it possible to model them as well as optimization[8].

The development of an accurate model for gloss change because of laser marking parameters variations is extremely complicated due to many process parameters involved. Variations in the quality of the marking process may be observe between processing cycles performed with the same laser equipment, and constant operating conditions and material properties. This poor reproducibility arises from the high sensitivity of the laser marking process to small changes in the operating parameters (such as laser power and etc), as well as to process disturbances (such as varying absorptivity, surface condition, geometry changes and workpiece thickness changes). Therefore, modeling of the laser marking parameters effect on the product's gloss is highly nonlinear and complicated.

In this paper, the effective parameters of the gloss in the laser marking process were modeled using adaptive neuro-fuzzy inference systems (ANFIS) and ANNs. Both models were trained with experimental data, namely samples that were obtained base on the 3^4 full factorial design.ANFIS was introduced to fuse fuzzy systems and neural networks into an integrated system to reap the benefits of two techniques [9]. It uses ANNs learning rule and can construct an input-output mapping for nonlinear systems. The ANFIS learns features in the data set and adjusts the system parameters according to a given error criterion [10-12]. Beside the ANFIS technique, there is a sparse use of ANNs in the different fields of science and engineering to model the complex and nonlinear problems. ANNs have a history of some six decades, but have found solid applications only in the past 20 years [13]. ANN is a type of artificial intelligence (AI) that mimics the behavior of human brain and is famous for its robustness (adaptability) due to the use of a generalization technique instead of memorization. Generally, an ANN is a highly interconnected network which consists of large number of processing elements (called neurons) and connections between them with coefficients (called weights) bound to connections in an architecture inspired by the brain which could learn the characteristic of patterns by examples [14].

The problems with ANFIS model are selecting input learning variables and choosing the optimum model parameters. To overcome the first problem, the method proposed by Jang [15] was utilized. Also two strategies were used for choosing the optimum parameters of ANFIS: Various kinds of membership functions with equal number of membership functions and various kinds of membership functions with different number of membership functions, which the latter one is on the basis of the analysis of variance (ANOVA), resulted from design of experiments (DOE) results whichis developed by authors. In the same manner, a difficult task in the building an ANN model is selection of optimum parameters such as number of neurons in each hidden layer, activation functions of neurons, training algorithms and network architectures. To cope with this problem, two different strategies were introduced: First strategy consisting of different neurons for several networks at the uniform threshold function for layers and subsequent strategy involve different neurons for several networks at the various threshold functions for layers.

Also, for performance measurement or model validation of evaluators, in addition to widely practiced criteria, a new method was proposed by authors which is base on the optimization results

Finally, the best predictor was introduced as the objective function for optimization process using GA and PSO, to obtain the maximum gloss of the gold. Contrary to PSO parameters selection, optimum selection of GA parameters is a difficult task and time consuming one. To conquer this problem, a new method isproposed by authors. In this method, the optimum number of population size is found and then other parameters are tuned base on the specific runs. In order to check the accuracy of calculations, the best Optimization result was taken as the optimum working parameters and was validated by a verifying experiment. Figure1 illustrates the usage of AI methodology for modeling and optimization of process.

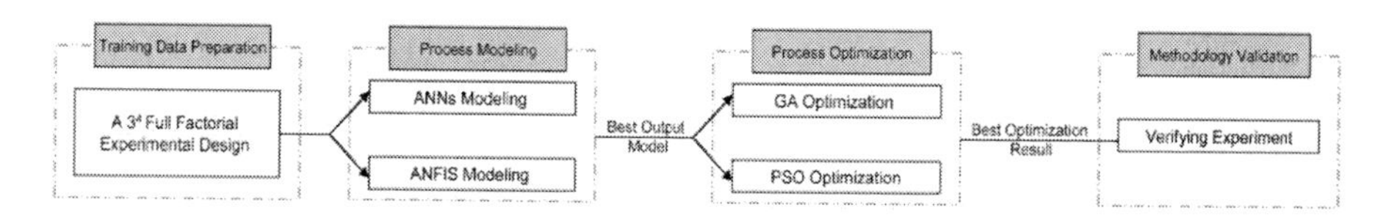

Figure 1.Application of AI techniques for modeling and optimization in this study.

The chapter is organized as follow. The fundamentals of ANFIS, ANNs, GA and PSO are presented in section 2. Section 3 illustrates experimental procedure; Section 4 describes implementation of ANFIS and ANN for modeling, Section 5 explains the GA and PSO implementation and a discussion about results; section 6 states the model validation step and section 7 concludes the chapter.

2. ANFIS, ANNs, GA and PSO

2.1. Adaptive Neuro-Fuzzy Inference System

In the modeling by fuzzy system (FS) concept, as shown in figure2, in general, the crisp input is converted into fuzzy by means of fuzzy sets and base on the fuzzification method. Fuzzy sets form the building blocks for fuzzy IF-THEN rules which have the general form of "IF x is A THEN y is B", where A and B are fuzzy sets. After fuzzification, the rule base is formed. The fuzzy inference engine (algorithm) combines fuzzy IF–THENrules into a mapping from fuzzy sets in the input space *X* to fuzzy sets in the output space *Y* based on the fuzzy logic principles. Finally, defuzzification is used to convert fuzzy value to the real world value which is the output [16].

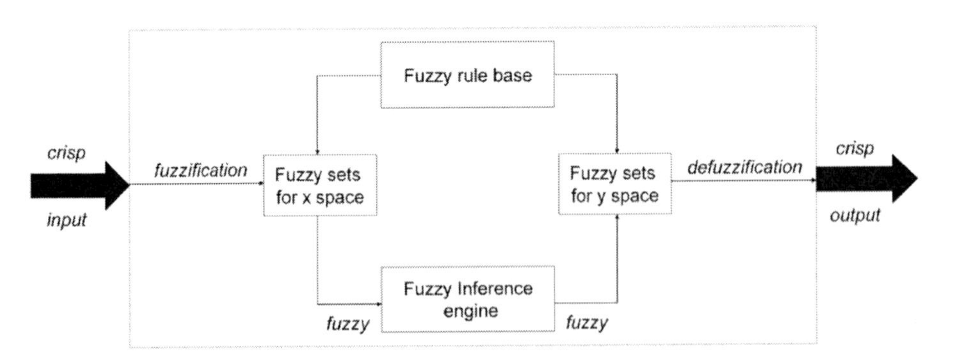

Figure 2.Modeling procedure by FS.

ANFIS represents a systematic way for implementing the aforesaid FS concept. It uses neural network approach for the solution of function approximation problems. Data driven procedures for the synthesis of ANFIS networks are typically based on the clustering a training set of numerical samples of the unknown function to be approximated. Like ANNs, ANFIS is model free predictor and has been successfully applied to different fields of science, engineering and various kinds of applications such as classification tasks, rule-based process controls, and pattern recognition problems and so on. This paper utilizes a fuzzy inference system comprises of the fuzzy model proposed by Takagi, Sugeno and Kang [17-18]to formalize a systematic approach to generate fuzzy rules from an input output data set.

2.1.1. Anfis Architecture

ANFIS structure consists of nodes and directional links through which the nodes are connected. Parts of the nodes are adaptive, which means each output of these nodes depends on the parameters pertaining to this node and the learning rule specifies how these parameters should be changed to minimize a prescribe error measure [19].

For simplicity, it is assumed that the fuzzy inference system under consideration has two inputs and one output. The rule base contains two fuzzy if-then rules of Takagi and Sugeno's type (18)as follows:

If x is A and y is B then z is $f(x, y)$

where A and B are the fuzzy sets in the antecedents and z = f (x, y) is a crisp function in the consequent. $f(x, y)$ is usually a polynomial for the input variables x and y. But it can also be any other function that can approximately describe the output of the system within the fuzzy region as specified by the antecedent. When $f(x, y)$ is a constant, a zero order Sugeno fuzzy model is formed, which may be considered to be a special case of Mamdani fuzzy inference system [20]that each rule consequent is specified by a fuzzy singleton. If $f(x, y)$ is taken to be a first order polynomial, a first order Sugeno fuzzy model is formed. Suppose that the rule base contains two fuzzy if then rules of first order Takagi and Sugeno's type, stated as below:

Rule 1 : If (x is A_1) and (y is B_1) then $(f_1 = p_1x + q_1y + r_1)$

Rule 2 : If (x is $A_{2)}$ and (y is B_2) then $(f_2 = p_2x + q_2y + r_2)$

Here type-3 fuzzy inference system proposed by Takagi and Sugeno[18] is used. In this inference system the output of each rule is a linear combination of input variables added by a constant term. The final output is the weighted average of each rule's output. The corresponding equivalent ANFIS structure is shown in figure3, where circles and squares indicate a fixed node and adaptive node, respectively.

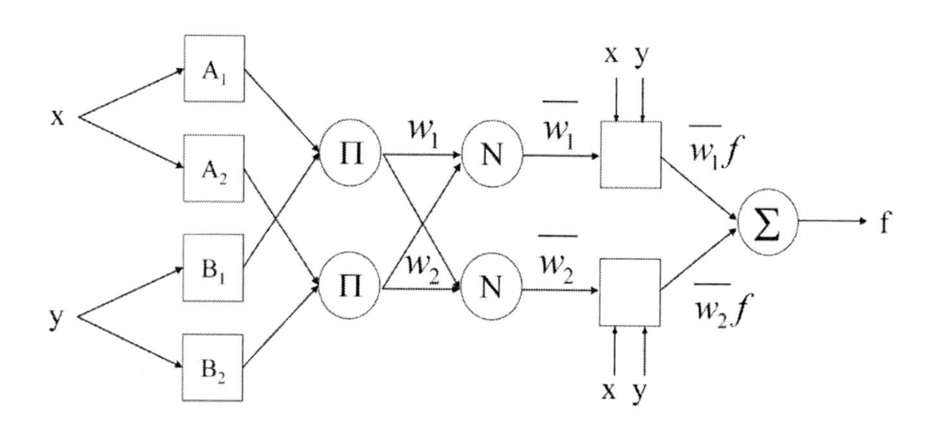

Figure 3.Architecture of ANFIS.

The node functions in the same layer are of the same function family as described below:

Layer 1: Every node i in this layer is a square node with a node function

$$O_1^i = \mu A_i(x) \tag{1}$$

where x is the input to node i and A_i the linguistic label (small, large, etc.) associated with this node function. Usually $\mu A_i(x)$ is selected to be bell shaped with maximum equal to 1 and minimum equal to 0, such as the generalized bell function:

$$\mu A_i(x) = \frac{1}{1+[(\frac{x-c_i}{a_i})^2]b_i} \tag{2}$$

Or the Gaussian function:

$$\mu A_i(x) = \exp[-(\frac{x-c_i}{a_i})^2] \tag{3}$$

where{a_i, b_i, c_i} are the parameter set. As the values of these parameters change, the bell-shaped functions vary accordingly, thus exhibiting various forms of membership functions (MFs) on linguistic label Ai. In fact, any continuous and piecewise differentiable functions, such as commonly used trapezoidal or triangular-shaped MFs, are also used for node functions in this layer. Parameters in this layer are referred to as premise parameters.

Layer 2: Every node in this layer is a circle node labeled Π, which multiplies the incoming signals and sends the product out. For instance,

$$w_i = \mu A_i(x)\mu B_i(x), i=1, 2... \tag{4}$$

Each node output represents the firing strength of a rule (In fact, other T-norm operators that performs generalized AND can be used as the node function in this layer).

Layer 3: Every node in this layer is a circle node labeled N. The *i*th node calculates the ratio of the *i*th rule's firing strength to the sum of all rules' firing strengths:

$$\overline{w_i} = \frac{w_i}{w_1 + w_2}, i=1, 2... \tag{5}$$

For convenience, outputs of this layer will be called normalized firing strengths.

Layer 4: Every node i in this layer is a square node with a node function

$$O_i^4 = w_i f_i = \overline{w_i}(p_i x + q_i y + r_i), \tag{6}$$

where w_i is the output of layer 3, and {p_i, q_i, r_i} is the parameter set. Parameters in this layer will be referred to as consequent parameters.

Layer 5: The single node in this layer is a circle node labeled $\sum$ that computes the overall output as the summation of all incoming signals, i.e.,

$$O_i^5 = \sum_i \bar{w}_i f_i = \frac{\sum_i w_i f_i}{\sum_i w_i} \tag{7}$$

Thus an adaptive network has been constructed.

2.1.2. ANFIS Learning Algorithm

The aim of the training algorithm for this architecture is tuning all the changeable parameters to make the ANFIS output match the training data. Note that here parameters $\{a_i, b_i, c_i\}$ of the membership function are fixed, and describe the sigma, slope and centre of the bell membership functions, respectively. Thus, the output of the ANFIS model can be written as:

$$f = \frac{w_1}{w_1 + w_2} f_1 + \frac{w_2}{w_1 + w_2} f_2 \tag{8}$$

Substituting Eq. (5) into Eq. (8) yields:

$$f = \bar{w}_1 f_1 + \bar{w}_2 f_2 \tag{9}$$

Substituting the fuzzy if–then rules into Eq. (9), it becomes:

$$f = \bar{w}_1 (p_1 x + q_1 y + r_1)_1 + \bar{w}_2 (p_2 x + q_2 y + r_2) \tag{10}$$

which is a linear combination of the changeable consequent parameters $\{p_1, q_1, r_1\}$ and $\{p_2, q_2, r_2\}$. The least squares method can be used to identify the optimal values of these parameters easily. When the premise parameters are not fixed, the search space becomes larger and the convergence of the training becomes slower. A hybrid algorithm combining the least squares method and the gradient descent method is adopted to solve this problem. The hybrid algorithm is composed of a forward pass and a backward pass. The least squares method (forward pass) is used to optimize the consequent parameters with the premise parameters fixed. When the optimal consequent parameters are found, the backward pass starts immediately. The gradient descent method (backward pass) is used to optimize the premise parameters corresponding to the fuzzy sets in the input domain. The output of the ANFIS is calculated by employing the consequent parameters found in the forward pass. The output error is used to adapt the premise parameters by means of a standard back-propagation algorithm. It has been proven that this hybrid algorithm is highly efficient in training the ANFIS[19].

2.2. Artificial Neural Networks

The past two decades have seen an explosion of renewed interest in the areas of Artificial Intelligence and Information Processing. Much of this interest has come about with the successful demonstration of real-world applications of ANNs and their ability to learn. Initially proposed during the 1950s, the technology suffered a roller coaster development accompanied by exaggerated claims of their virtues, excessive competition between rival research groups, and the perils of boom and bust research funding. ANNs have only recently found a reasonable degree of respectability as a tool suitable for achieving a nonlinear mapping between an input and output space. ANNs have proved particularly valuable for applications where the input data set is of poor quality and not well characterized. The number of types of ANNs and their uses is very high. Since the first neural model by McCulloch and Pittsthere have been developed hundreds of different models considered as ANNs. The differences in them might be the functions, the topology, the learning algorithms and so forth which amongst them, backpropagation networks have found most applications in the modeling and many other networks are based on it. Such a network learns using the backpropagation algorithm for discovering the appropriate weights. In this research, the backpropagation networks were utilized, namely feed forward and cascade forward networks. Furthermore, a Levenberg-Marquardt (LM) learning algorithm was utilized for tuning of the network weights. In addition, the logistic sigmoid (LOGSIG) and tangent sigmoid (TANSIG) activation functions in two hidden layers were used.

2.2.1. Network Types

(a) Feed Forward Back Propagation (FFBP). This network consists of one input layer, one or several hidden layers and one output layer. Usually back propagation learning algorithm (BP) is used to train this network. In the case of BP algorithm, first the output layer weights are updated. For each neuron of the output layer, a desired value exists. By this value and the learning rules, weight coefficient is updated. For some problems, the BP algorithm presents suitable results, while it ends to improper results for others. In some cases, the learning process is upset as a result of getting trapped in local minimum. This is because of the answer, lying at the smooth part of the threshold function. Figure4 shows the training process of BP algorithm for updating weights and biases.

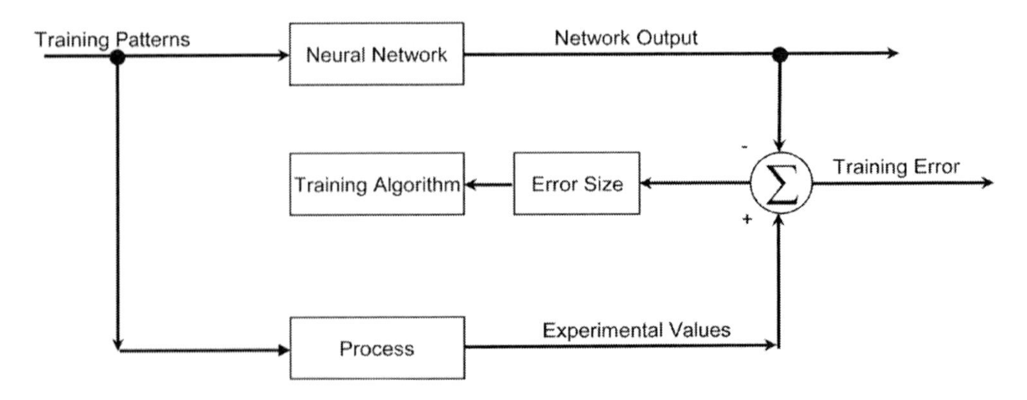

Figure 4. Training process of the back propagation networks.

During training of this network, calculations are done from input of network toward output, and then values of error propagate to prior layers. Output calculations are done layer to layer and output of each layer would be input of next layer.

(b) Cascade Forward Back Propagation (CFBP). This network like FFBP network uses the BP algorithm for updating weights, but the main symptom of this network is that each layer's neurons are related to all previous layer neurons.

2.2.2. Training Algorithm

Levenberg-Marquardt algorithm (LM). Gradient-based training algorithms, such as back propagation, are most commonly used by researchers. They are not efficient because the gradient vanishes at the solution. Hessian based algorithms allow the network to learn features of a complicated mapping more suitable. The training process converges quickly as the solution is approached, because the Hessian does not vanish at the solution. To benefit from the advantages of Hessian based training, we use Levenberg-Marquardt algorithm. The LM algorithm is a Hessian based algorithm for non-linear least squares optimization. The steps involved in training a neural network in batch mode using the LM algorithm are as follows [21]:

1. Presenting all inputs to the network and computing the corresponding network outputs and errors.
2. Computing the jacobian matrix, J; where x is weights and biases of the network.
3. Solving the LM weight update equation

$$x_{k+1} = x_k - [J^T J + \mu I]^{-1} J^T e \tag{11}$$

where x_k is calculated weight in the prior step, x_{k+1} is weight value in new step, μ is training parameter, $J^T J$ is Hessian matrix , I is the identity matrix and $J^T e$ is gradient (which e is the network error vector).

4. Recomputing the error using x +Δx. If this new error is smaller than that computed in step 1, then reduce the training parameter μ by μ– let x= x+Δx, and go back to step 3. μ+ and μ– are predefined values set by the user.

2.3. Genetic Algorithm

Genetic algorithm is a computerized search and optimization algorithm based on the mechanics of natural genetics and natural selection. Prof. Holland of MichiganUniversity envisaged the concept of these algorithms in the mid-sixties. GA is good at taking larger, potentially huge, search spaces and navigating them looking for optimal combinations of things and solutions which we might not find in a lifetime.

A difference between genetic algorithms and most of the traditional optimization methods is that GA uses a population of points at one time in contrast to the single point approach by traditional optimization methods[22].This means that GA processes a number of designs at the same time. GA uses randomized operators instead of transition rules in traditional optimization methods. GA is different from more normal optimization and search procedures in four ways:

- GA work with a coding of the parameter set, not the parameters themselves.
- GA search from a population of points, not a single point.
- GA use probabilistic transition rules, not deterministic rules.
- GA use objective function information, not derivatives or other auxiliary knowledge.

Taken together, these differences - direct use of a coding, search from a population, blindness to auxiliary information and randomized operators-contribute to genetic algorithm robustness and resulting advantage over other more commonly used techniques.

The mechanics of Genetic Algorithm is simple, involving three stages, (a) population initialization (b) operators (c) chromosome evaluation

(a) Population Initialization

The generation of the initial population in GA is usually done randomly; however, the initial population must consist of strings of valid sequences, satisfying all precedence relations.

(b) Operators

I) Selection, the selecting of selection operators is an important part in GA. There are many different selection operators presented by some researchers. For example, the ''elitist model'' and the ''expected value model'' were presented by De Jong [23] or the ''tournament selection'' was presented by Goldberg et al [24].

II) Reproduction, this is the second of the genetic operators. It is a process in which copies of the strings are copied into a separate string called the 'mating pool', in proportion to their fitness values. This implies that strings with higher fitness values will have a higher probability of contributing more strings as the search progresses.

III) Crossover, this operator is mostly responsible for the progress of the search. It swaps the parent strings partially, causing offspring to be generated. In this, a crossover site along the length of the string is selected randomly, and the portions of the strings beyond the crossover site are swapped.

IV) Mutation, although the processes of reproduction and crossover produce many new chromosomes, they do not introduce any new information into a population at the bit level. Mutation is a random occasional alteration of the information contained in the chromosomes. Genes occasionally mutate to become new genes with a low probability.

Therefore, in order to mimic the process of evolution, the process of mutation introduces bits to mutate with a low probability by inverting a randomly chosen bit onto a chromosome.

V) Migration, migration is the movement of individuals between subpopulations. Every so often, the best individuals from one subpopulation replace the worst individuals in another subpopulation. Migration parameters are direction, fraction and interval.

(c) Chromosome Evaluation

When all the individuals (process plans) in the population have been determined to be feasible, i.e. an operation precedence is guaranteed, they can be evaluated based on the fitness functions.

2.4. Particle Swarm Optimization

PSO is a population-based optimization technique proposed firstly by Kennedy and Eberhart in [25] for the unconstrained minimization problem. Particle swarm optimization (PSO) is inspired from studies of various animal groups and it has been proven as a powerful competitor to the other evolutionary algorithms such as genetic algorithm [25].

PSO system combines local search method (through self experience) with global search methods (through neighboring experience), attempting to balance exploration and exploitation. PSO is widely used to solve nonlinear and multi-objective problems such as optimization of weights of neural networks (NN), electrical utility, computer games, and mobile robot path planning, etc.[26-28].

In a PSO system, multiple candidate solutions coexist and collaborate simultaneously. Each solution called a ‘‘particle’’, flies in the problem search space looking for the optimal position to land. A particle, as time passes through its quest, adjusts its position according to its own ‘‘experience’’ as well as the experience of neighboring particles. Tracking and memorizing the best position encountered build particle’s experience. For that reason, PSO possesses a memory (i.e. every particle remembers the best position it reached during the past). PSO system combines local search method (through self experience) with global search methods (through neighboring experience), attempting to balance exploration and exploitation. A particle status on the search space is characterized by two factors: its position and velocity, which are updated by following equations.

$$V_i(t+1)=\omega V_i(t)+c_1r_1[p_{best_i}-X_i(t)]+c_2r_2[g_{best_i}-X_i(t)] \tag{12}$$

$$X_i(t+1)=X_i(t)+V_i(t+1) \tag{13}$$

where *Vi* = *[vi,1, vi,2, . . ., vi,n]* called the velocity for particle *i*, which represents the distance to be traveled by this particle from its current position; *Xi* = *[xi,1,xi,2, . . .,xi,n]* represents the position of particle *i*; p_{best_i} represents the best previous position of particle *i* (i.e. local-best position or its experience); g_{best_i} represents the best position among all particles in the population *X* = *[X1,X2, . . .,XN]* (i.e. global-best position); r_1 and r_2 are two independently uniformly distributed random variables with range *[0, 1]*; c_1 and c_2 are positive constant parameters called acceleration coefficients which control the maximum step size; ω is called the inertia weight that controls the impact of previous velocity of particle on its current one. In the standard PSO, Eq. (12) is used to calculate the new velocity according to its previous velocity and to the distance of its current position from both its own best historical position and its neighbors’ best position. Generally, the value of each component in

Vi can be clamped to the range *[v_{min}, v_{max}]* to control excessive roaming of particles outside the search space. Then the particle flies toward a new position according Eq. (13). This process is repeated until a user-defined stopping criterion is reached. The standard PSO is illustrated in figure5, where *N* denotes the size of population, *fi* represents the function value of the *ith* particle, and *fbest[i]* represents the local-best function value for the best position visited by the *ith* particle. For interested readers could refer to [25, 29] for more details and modifications on PSO.

Step 1 (Initialization): For each particle i in the population:

Step 1.1: initialize Xi randomly.
Step 1.2: initialize Vi randomly.
Step 1.3: evaluate fi.
Step 1.4: initialize Pi with the index of the particle with the best function value among the population.
Step 1.5: initialize Pi with a copy of Xi , .

Step 2: Repeat until a stopping criterion is satisfied:
Step 2.1: find Pg such that .
Step 2.2: for each particle i, Pi = Xi
If
Step 2.3: for each particle i, update Vi and Xi according to equation 7 and 8.
Step 2.4: evaluate fi for all particles.

Figure 5. Standard PSO.

3. Input/Output Variables

An Nd: YAG laser (ACI-Magic Mark/5Watt, 1064nm) was used to mark the samples on the gold. All of the samples were marked in the same way, the circles with 6mm diameters. The experiments were done on the gold plates which was 750-carat gold. In order to avoid producing any curvy surface, the gold surface was polished and completely leveled using a press, because even a very small bulge in the surface may result in the deviation in the course of reflected rays. Samples were obtained base on the 3^4 full factorial design. Table 1 shows process factors and their respective level.

Table 1. Factors and levels used in the 3^4 factorial design study

Factors	Levels		
	-1	0	+1
Power Percent (5 watt) (X1)	97	98.5	100
Speed (mm/s) (X2)	100	150	200
QSF (Hz) (X3)	8000	12000	16000
PW (µm) (X4)	5	15	40

The gloss values were measured with a glossmeter (NOVA 407, elcometer Ltd). The measurement of the gloss value R_s is a relative measurement [6]. The measurement results are related to a highly polished, black glass standard with a defined refractive index of 1.567. The black glass standard has an assigned specular gloss value of 100 (calibration). In order to obtain comparable results, the measurement device and operation have been defined in international specifications: ISO 2813, ASTM D 523, JIS Z 8741, and DIN 67530 [30-33]. The glossmeter was calibrated before each recording session with mentioned standard sample supplied by the manufacturer. The angle of illumination highly influences the measurement results. Due to highly reflective feature of the gold, in this study the measurements were obtained at illumination angle of 20^0[6]. The data set in table 2 presents approximately 20% of 81 data sets corresponding to samples and operational parameters.

Table 2. 20% of 81 data sets corresponding to samples and parameters of laser marking

Pattern	Parameters				Target
	Power	Speed	QSF	P-W	
	X1	X2	X3	X4	20
1	97	100	8000	5	616
2	97	100	8000	15	601
13	97	150	12000	5	799
14	97	150	12000	15	812
17	97	150	16000	15	665
27	97	200	16000	40	756
30	98.5	100	8000	40	792
34	98.5	100	16000	5	634
35	98.5	100	16000	15	796
42	98.5	150	12000	40	409
46	98.5	200	8000	5	752
47	98.5	200	8000	15	771
48	98.5	200	8000	40	793
56	100	100	8000	15	693
65	100	150	8000	15	521
73	100	200	8000	5	477

4. ANFIS AND ANNSIMPLEMENTATION

4.1.Model Building Methodology

Construction of a model for a given data set can be categorized in 3 steps as shown in figure6.

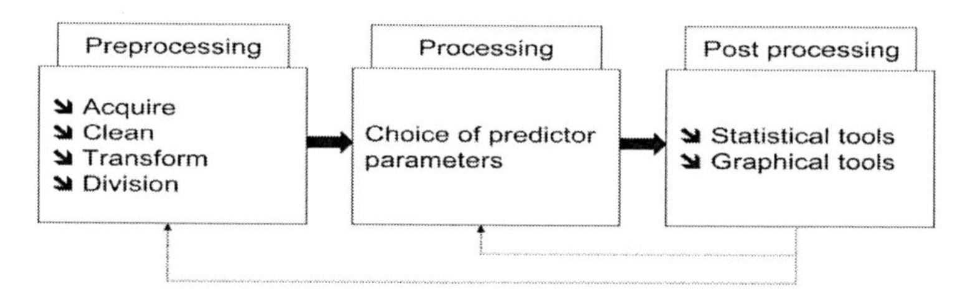

Figure 6. Model building strategy.

(a) First step belongs to preprocessing of data: collection of sufficient data to represent all underlying patterns (acquire); removing obvious errors and extraneous outliers (clean); normalization of data (transform); separation of data into training, validation and test sets (division). Each sub step has significant effect in the model building procedure. In the acquiring, obtained data should explore the domain of the system as far as possible. Poor data collection may result in poor performance of evaluator. In addition, it is important to spend time for cleaning the data. There are potential errors in the data which should be removed (see figure 7). This sub step reduces the domain of modeling problem, but it increases the accuracy of the model (a comparison of some of ANFIS modeling and ANN modeling results before and after cleaning of data were given in appendix A). Furthermore, to avoid introducing unnecessary bias resulting from variables being measured on the different scales, data should be normalized prior to processing step. Moreover, transformation increases the training velocity. Finally, all of data should be separated into train, validation and test sets, in order to train, measuring the generalization and providing the independent performance measurement, respectively[34-35]. It is worthy of note that ANFIS is so sensitive to input variables for training and this sub step has significant role in the neuro-fuzzy modeling approach.

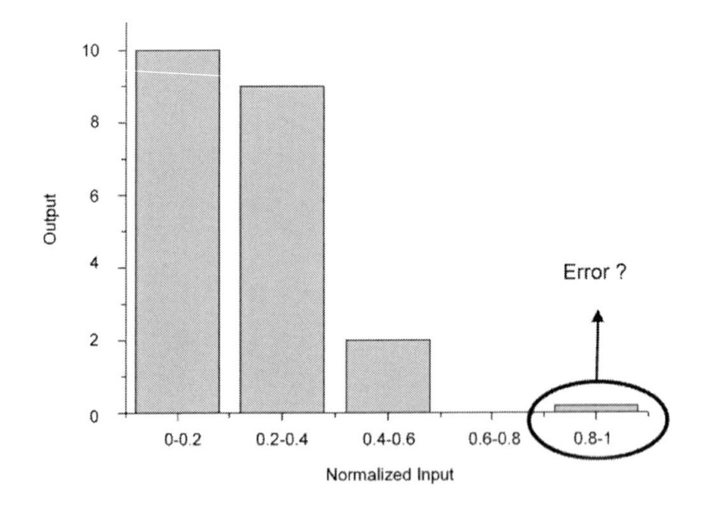

Figure 7. Potential errors in the obtained data.

(b) Not only does processing step specify the evaluator parameters settings, but it also defines how to establish balance between overtraining and poor learning. ANFIS parameters include training algorithm type, number of MFs, MFs type and training epoch. ANN has more parameters settings, such as network type, training algorithms, number of neurons in the hidden layers, activation functions of neurons, etc.

(c) Final step, post processing, is validation of the candidate evaluator/evaluators. Model validation or performance measurement is an important step amongst the model building sequences. There are two different approaches for the model validation: statistical (numerical) methods such as determination coefficient (R^2), root mean square error (RMSE), mean absolute error (MAE) and so forth and graphical methods such as residual scatter plots and so on. In this paper, we used determination coefficient R^2 and *RMSE* according to the following equations as a yardstick for performance measurement.

$$RMSE = \sqrt{\frac{1}{T}\sum_{k=1}^{T}(S_k - T_k)^2} \quad (14)$$

where T denotes the number of data patterns, S_k is network output of *k*th pattern and T_k is target output of*k*th pattern (experimental output data).

$$R^2 = 1 - \frac{\sum_{k=1}^{T}[S_k - T_k]}{\sum_{k=1}^{T}[S_k - T_m]}, \quad T_m = \frac{\sum_{k=1}^{T} S_k}{T} \quad (15)$$

When the *RMSE* is at the minimum and R^2 is high, ($R^2 \geq 0.8$); a model can be judged as very good (14). Moreover, the fact that the *RMSE* error values are very closes to each other, is an indication of a sufficient regularization [36].

Note that in the figure6, we have given feedbacks from third step to first and second steps.It means that if not all the designed models satisfy conditions of the third step, the designer should revise the preprocessing and processing settings, i.e. changes division and distribution of data sets, normalization, cleaning, changing the ANFIS or ANN parameters settings, etc.

Additionally, in this paper a new method for performance measurement of evaluators is proposed. The proposed model validation method is based on the optimization result. Suppose that for a highly nonlinear physical system we can have a rough estimate about its maximum or minimum value. Besides, suppose for such highly nonlinear physical system, there is a unique experimental model that has the best prediction outputs. Now, if one uses this model as an evaluator for an optimization technique, it is expected that the best optimization result will have been obtained. For instance, if the optimization problem is maximization, it is expected that the maximum result for the mentioned system will have been obtained which this result is approximate to the rough estimate result. In the other words, the more the evaluator is accurate, the more the result is approximate. Conversely, the more the evaluator is inaccurate, the more the result is far away from the rough estimate. This assumption is heuristically reasonable. Furthermore, this method is not a final criterion, but it can be used as an assistant yardstick, in addition to the other criteria.

For our system in this paper, we expected that the best maximum value of gloss is 1200 gloss, because in the best condition of a laser marked gold, the glossmeter will be show the gloss value of 1200, the maximum possible value for a highly gloss surface.

4.2. ANFIS Modeling

Three steps of model building for ANFIS will be discussed as following:

(a) Preprocessing.Data acquisition was performed base on the 3^4 full factorial design, as noted before. After cleaning sub step, they were reduced to 64 data sets. For transformation, min-max method was used to normalize parameters data (input data) into boundary of [0-1]. Furthermore, the target data (output data) were normalized into boundary of [0.601-0.812] using decimal scaling [35], according to the following equations:

$$\text{Min-max scaling: } X_n = \frac{X_i - X_{min}}{X_{max} - X_{min}} \tag{16}$$

in which X_n is normalized parameter, X_i is real parameter, X_{min} is minimum parameters value (of each X_n) and X_{max} is maximum parameters value (of each X_n).

$$\text{Decimal scaling: } Y' = \frac{Y}{10^n} \tag{17}$$

where Y' is normalized target data, Y is real data and we determine n=3 to proportionate normalization boundary of parameters and target data.

Table 3. RMSE of the first train of different sets and corresponding division fractions

Division Fraction		
70-30	60-40	50-50
RMSE	RMSE	RMSE
0.0513	0.0449	0.0346
0.0472	0.0477	0.039
0.0493	0.0509	0.0455
0.0426	0.0448	0.0403
0.0393	0.0493	0.0455
0.0452	0.0365	0.0415
0.0436	0.0506	0.0319
0.0504	0.0292	0.029
0.0433	0.0428	0.0229
0.0478	0.0432	0.0425
0.0411	0.0458	0.0368
0.0489	0.0464	0.0199
0.0391	0.0357	0.0406
0.0416	0.039	0.0515
0.054	0.0269	0.0204
0.0415	0.0447	0.0221
0.0434	0.0288	0.0288
0.0357	0.0412	0.0368
0.0335	0.0367	0.0392
0.0396	0.0511	0.0299

One of the great challenges in the modeling of nonlinear systems (specifically using ANFIS) is selecting the important input variables from all possible input variables. That is, it is necessary to do input selection that finds the priority of each candidate inputs and uses them accordingly [15, 37]. This concept relates to the division sub step of preprocessing step. Base on the author's experience [38], both division fraction of train, validation and test sets and distribution of data into these sets, are effective factors. In this research study, in order to do input selection for ANFIS learning, Jang's method [15]is utilized and also, it is expanded for determining the optimum division fraction. The utilized input selection method is on the basis of assumption that "the ANFIS model with the smallest root mean squared error (RMSE) after one epoch of training has a greater potential of achieving a lower RMSE when given more epochs of training" [15]. We expanded this method and added the term of "division fraction" to it. Using a written code in the MATLAB, we made different sets of train/validation/test sets with various division fractions and then trained them for one epoch. The smallest one was selected as the best data set for training. Table 3 summarizes some results of the program.

In this table, 70-30 denotes 70 percent of data were selected for train and remained for validation, test, and so on. In addition, RMSE implies the RMSE of sets only after one epoch of training. The set with 0.0199 RMSE at the first epoch was selected for processing step.

(b) Processing. Fuzzy Logic Toolbox of MATLAB was used for this part of the research. In order to find the optimum ANFIS parameters, the methodology according to [38] is used. Two different strategies were utilized: various kinds of membership functions with equal number of membership functions and various kinds of membership functions with different number of membership functions. The latter strategy is on the basis of the ANOVA, resulted from DOE analysis. The proposed method offers that we choose more MFs for significant parameters.

Table 4. ANOVA resulted from DOE analysis

Source	DF	SS	MS	F	*p*
Power	2	14693	73466	4.69	0.012
Speed	2	97647	48823	3.12	0.050
QSF	2	41386	20693	1.32	0.273
PW	2	47864	23932	1.53	0.224

ANOVA table includes DF, representing the degrees of freedom from each source (the number of levels for the factor minus one); SS, demonstrating the sum of squares between groups (factors), MS that stand for mean squares which are found by dividing the sum of squares by the degrees of freedom, F which is calculated by dividing the factor MS by the error MS to determine whether a factor is significant and p-value to determination whether a factor is significant; typically by comparing against an alpha value of 0.05. If the p-value is lower than 0.05, then the factor is significant. In other words, larger the magnitude of Fand smaller the value of p, the more significant is the corresponding parameter [39-40]. According to this table, in the second strategy, more MFs were selected for laser power (X1) and speed (X2).

Since it has been proven that hybrid algorithm is highly efficient in the training of ANFIS [19], hybrid method was selected as the learning algorithm. Besides, Gaussian (gaussmf) and bell shape (gbell) MFs were selected as MFs types.

(c) Post processing.Results of two strategies were shown in table 5 and table 6.

Table 5. Various kinds of MFs with equal number of MFs

MFs Type	MFs Number	Train		Validation		Test	
		R^2	RMSE	R^2	RMSE	R^2	RMSE
gbell	2 2 2 2	0.8746	0.0178	0.0321	0.0966	0.1964	0.0682
	3 3 3 3	1	0	0.00002	0.4459	0.0618	0.4658
gaussmf	2 2 2 2	0.8735	0.018	0.0322	0.0957	0.1916	0.0683
	3 3 3 3	1	0	0.0074	0.3121	0.0564	0.518

Table 6. Various kinds of membership functions with different number of membership functions

MFs Type	MFs Number	Train		Validation		Test	
		R^2	RMSE	R^2	RMSE	R^2	RMSE
gbell	3 3 2 2	0.9288	0.0135	0.0263	0.1429	0.5618	0.0937
gaussmf	3 3 2 2	0.9256	0.0138	0.0155	0.1522	0.4131	0.1157

The best result of the first strategy is related to the bell shape (gbell) MFs with two MFs for each input that has RMSE=0.0178 andR^2=0.8746. It is seen that increasing the MFs to three MFs for each input leads to overtraining of the model. Furthermore, it can be concluded that gbell MFs has better result compared to the Gaussian one. The best result for the second strategy belongs to thegbell MFs with RMSE=0.0135 and R^2=0.9288. Thus, the best ANFIS model is related to the second strategy.

4.3. ANNs Modeling

Three steps of model building for ANN will be discussed as following:

(a) Preprocessing.This step For ANN is like ANFIS preprocessing step, except that parameters were normalized into the boundary of [-1,1] using the following equation:

$$\text{Min-max scaling: } X_n = \frac{X_i - X_{\min}}{X_{\max} - X_{\min}} \times (2) + (-1) \tag{18}$$

(b) Processing. Considering and applying four inputs in all experiments, the gloss value derived from different conditioning with these conditions, networks with four neurons in input layer (laser power, speed, QSF and PW) and one neuron in output layer (gloss) and two hidden layers was designed. Figure8 shows the considered neural network topology, Input and output parameters.

Neural network toolbox of MATLAB software was used for this part of the research. The MATLAB neural networks toolbox provides a complete set of functions and a graphical user interface for the design, implementation, visualization and simulation of neural networks. It supports the most commonly used supervised and unsupervised network architectures and a comprehensive set of training and learning functions [41].

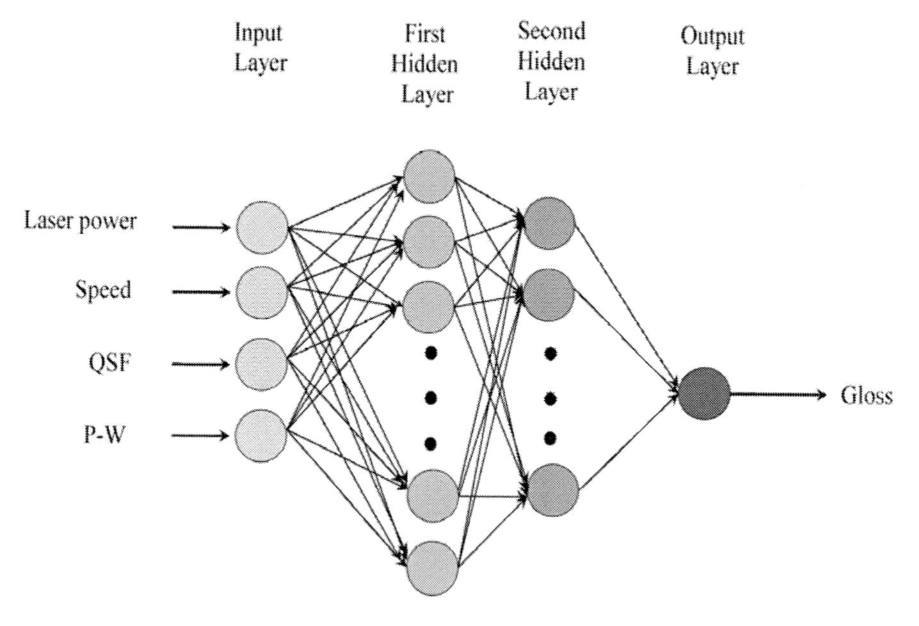

Figure 8. Considered ANN topology.

Two different strategies base on the [42]were utilized for ANNs optimum parameter settings:

- Different neurons for several networks at the uniform threshold function for layers and
- Different neurons for several networks at the various threshold functions for layers.

For obtaining desired answer, two networks were utilized; FFBP and CFBP neural networks. Training process by these networks is repetitive, when the error between the desired value and predicted value becomes minimum, training process towards to stability.

There is no formula to estimate the number of data points required training an ANN and the number can vary greatly depending on the complexity of the problem and the quality of the data, but many ANNs have been successfully trained with the smaller number of data points than the number of weights [43]. On the other hand, there are results available in the literatures that indirectly help in the selection of number of hidden nodes for a given number of training patterns and besides, there are formulas available in the manuscripts base on the rule of thumb describing a rough estimate about deciding the number of hidden nodes.

The increasing method, moreover, gives a good procedure for solving aforesaid problem. By this method, when the network traps into the local minimum, new neurons are added to network gradually. This method has the more practical potential for detecting the optimum size of network. Advantages of this method are:

- Network complexity increases gradually by increasing of neurons.
- The optimum size of network is always obtained by adjustments.
- Monitoring and evaluating local minimum is done during the training process.

In this work, a hybrid method, a formula and increasing method combination is proposed for setting the appropriate ANN parameters. We started with a number of hidden nodes resulted from the following formula and then continued by increasing method.

$$Q \geq \frac{N}{\varepsilon} \tag{19}$$

where Q is the number of training set, N is the number of adjustable parameters (weights) and ε is learning error percent [44]. Choosing $\varepsilon = 0.1$ and Q=49, the number of adjustable parameters is approximately obtained as 4.

Various threshold functions were used to obtain the optimized status:

$$\text{LOGSIG: } Y_j = \frac{1}{1 + exp(-X_j)} \tag{20}$$

$$\text{TANSIG: } Y_j = \frac{2}{(1 + exp(-2X_j)) - 1} \tag{21}$$

That $X_j$is the sum of weighted inputs for each neuron in jth layer and is computed as below:

$$X_j = \sum_{i=1}^{m} W_{ij} \times Y_i + b_j \tag{22}$$

in which$m$ is the number of output layer neurons, $W_{ij}$is the weight between ith and jth layers, $Y_i$is ith neuron output andb_j is bias of jth neuron for FFBP and CFBP networks. About 75% of all data randomly were selected for training of network with suitable topology and training algorithms. Training data are presented to the network during training, and the network is adjusted according to its error. Remained data are used for validation and test. Validation data are used to measure network generalization and to halt training when generalization stops improving. Testing data have no effect on training and provide an independent measure of network performance during and after the training.

(c) Post processing.Both of two strategies were used for FFBP and CFBP networks with learning algorithm of LM. The best results of used networks and algorithms are shown in the table 7 and table 8.

Table 7. Training algorithm for different neurons of hidden layers and several networks at the uniform threshold functions for layers

Network	Training Algorithm	Threshold Function	No. of Hidden Layers Neurons	Train R^2	Train RMSE	Validation R^2	Validation RMSE	Test R^2	Test RMSE
CFBP	LM	TANSIG	2-2	0.7455	0.0311	0.2281	0.0574	0.8837	0.0239
		LOGSIG	3-1	0.7813	0.0288	0.331	0.0429	0.7768	0.0546
		TANSIG	3-2	0.9805	0.5960	0.0086	0.153	0.9146	0.0307
		LOGSIG	3-3	0.9539	0.0133	0.6909	0.037	0.9333	0.0204
		TANSIG	4-2	0.8297	0.0259	0.5693	0.0399	0.7592	0.0413
		LOGSIG	4-4	0.9117	0.0184	0.4142	0.0511	0.7493	0.0778
		TANSIG	5-3	0.9537	0.0133	0.6362	0.041	0.8225	0.0591
		LOGSIG	6-3	0.9951	0.0049	0.2084	0.0685	0.8246	0.0312
		TANSIG	6-6	0.8774	0.0216	0.8496	0.0322	0.8617	0.1107
		TANSIG	9-8	0.992	0.0057	0.8253	0.0273	0.8043	0.0459
		LOGSIG	10-8	0.8971	0.0219	0.7174	0.0384	0.8518	0.0263
FFBP	LM	TANSIG	2-2	0.4303	0.0465	0.2034	0.0591	0.2731	0.1655
		LOGSIG	3-1	0.5029	0.043	0.5379	0.0439	0.3237	0.1281
		TANSIG	3-2	0.6787	0.0351	0.6759	0.041	0.6579	0.0614
		LOGSIG	3-3	0.7307	0.032	0.533	0.043	0.5504	0.0945
		TANSIG	4-2	0.7008	0.0338	0.1862	0.0561	0.5608	0.0719
		LOGSIG	4-4	0.8271	0.0258	0.6878	0.0364	0.6173	0.0794
		TANSIG	5-3	0.7303	0.0323	0.5473	0.0484	0.6001	0.1794
		LOGSIG	6-3	0.9395	0.0152	0.7404	0.0303	0.807	0.0658
		TANSIG	6-6	0.87	0.0221	0.8319	0.0378	0.8771	0.0257
		TANSIG	9-8	0.9468	0.0142	0.8719	0.028	0.8383	0.0579
		LOGSIG	11-9	0.9787	0.0097	0.8388	0.0247	0.8527	0.0383

Table 8. Training algorithm for different neurons of hidden layers and several networks at the various threshold functions for layers

Network	Training Algorithm	Threshold Function	No. of Hidden Layers Neurons	Train R^2	Train RMSE	Validation R^2	Validation RMSE	Test R^2	Test RMSE
CFBP	LM	T-T-L*	2-2	0.7089	0.0332	0.1819	0.0546	0.7434	0.0431
		T-L-T	3-1	0.7912	0.0282	0.6254	0.0424	0.6337	0.0629
		T-L-L	3-2	0.7983	0.0278	0.6297	0.0356	0.7276	0.0687
		L-T-T	3-3	0.9006	0.0202	0.212	0.0704	0.8831	0.0243
		L-T-L	4-2	0.9191	0.0175	0.192	0.0796	0.7641	0.1078
		L-L-T	4-4	0.9203	0.0176	0.4615	0.0592	0.7112	0.0409
		T-T-L	5-3	0.897	0.0199	0.3546	0.0584	0.8051	0.0698
		T-L-T	6-3	0.886	0.0208	0.0818	0.0616	0.8227	0.1406
		T-L-L	6-6	0.9316	0.0161	0.8295	0.0442	0.9433	0.1951
		T-T-L	9-8	0.9189	0.0193	0.8315	0.0398	0.8988	0.0252
		L-T-L	10-8	0.9813	0.0087	0.7532	0.0314	0.7783	0.0641
FFBP	LM	L-T-T	2-2	0.6319	0.0373	0.5284	0.0418	0.6234	0.0938
		L-T-L	3-1	0.6067	0.0390	0.7238	0.0345	0.2886	0.0528
		L-L-T	3-2	0.821	0.026	0.6722	0.0373	0.1907	0.0883
		T-T-L	3-3	0.8027	0.0271	0.624	0.0367	0.6679	0.0604
		T-L-T	4-2	0.8014	0.0275	0.6314	0.0391	0.232	0.086
		T-L-L	4-4	0.8182	0.0263	0.7397	0.0308	0.6339	0.388
		L-T-T	5-3	0.8391	0.0249	0.4472	0.047	0.6911	0.0788
		L-T-L	6-3	0.9095	0.0185	0.6275	0.0397	0.6042	0.0725
		L-L-T	6-6	0.9555	0.0131	0.8654	0.0239	0.872	0.057
		L-T-L	9-8	0.9525	0.0152	0.8889	0.0361	0.8667	0.052
		T-T-L	11-9	0.9524	0.014	0.8345	0.0438	0.8425	0.0513

*T and L stand for TANSIG and LOGSIG respectively.

In the first look, the best results of CFBP network with LM algorithm in the first strategy is related to 4-9-8-1 topology that produces*RMSE*=0.0057 and R^2=0.992 for the train set. The results for FFBP network in the first strategy shows the LOGSIG threshold function has better performance relative to TANSIG threshold function and 4-11-9-1 topology has better result.The best result for the second strategy for CFBP networks is for 4-6-6-1 topology and TANSIG-LOGSIG-LOGSIG threshold functions that has *RMSE*=0.0161 and R^2=0.9316. For FFBP network withthe second strategy, the network with 4-9-8-1 topology and threshold functions of LOGSIG-TANSIG-LOGSIG produces*RMSE*=0.0152 and R^2=0.9525 and has the suit performance.

4.4. Results and Discussion

In this study, the neuro-fuzzy computing technique and ANNs were used to evaluate the gloss of the laser marked gold due to the different laser marking process parameters. For performance measurement (model validation) of each method and judging amongst them, following criteria (filters) were utilized step by step:

- HighR^2 value ($R^2 \geq 0.8$) for train, validation and test sets [14].
- The more approximate optimization result value to the maximum expected gloss value (1200), according to the proposed method.
- Close RMSE value of train, validation and test set to each other (an indication of a sufficient regularization [36]).
- In almost equal conditions, the model with lower complexity is superior to the other ones.

In the ANFIS modeling, second strategy has better results compared to the first one and it is seen thatincreasing the MFs to three one for each input, causes overtraining of the model. Although gbell MFs with 3-3-2-2 number of MFs for power, speed, QSF and P-W respectively, have best result amongst other ANFIS model, but it doesn't satisfy a high R^2 value ($R^2 \geq 0.8$) for validation and test set.

Table 9. Training algorithm for different neurons of hidden layers and several networks at the various threshold functions for layers

Network	Threshold Function	No. of Hidden Layers Neurons	GA Output
CFBP	T-T-T	6-6	911
	T-T-T	9-8	937
FFBP	T-T-T	6-6	875
	T-T-T	9-8	837
	L-L-L	11-9	888
CFBP	T-L-L	6-6	961
	T-T-L	9-8	934
FFBP	L-L-T	6-6	849
	L-T-L	9-8	966
	T-T-L	11-9	850

In the ANN modeling, a similar situation can be seen which increasing the number of hidden nodes leads to overtraining ofthe model. There are ten different models satisfying the high R^2 value condition; that is, they are passing the first filter.

At this step, second criterion was applied. GA optimization method was used as the optimization technique and equal parameters settings were considered for all of evaluators. Table 9 demonstrates the results of GA optimization for aforesaid evaluators.

Using second filter, 5 evaluators which had approximate high values of GA outputs were passed from this step.

Using third criterion and attentive to the previous criteria, it is seen that CFBP network with 4-9-8-1 topology and TANSIG-TANSIG-LOGSIG threshold functions has the best results in comparison with other candidate networks, because it has more closer RMSE value of train, validation and test sets together, compared to the other networks.Figure 9(a) to figure 9(c) show the determination coefficient (R^2) value of the train, validation and test sets for this evaluator, respectively.

5. GA and PSO Implementation

5.1. Optimization

The goal of optimization is to find values of variables that minimize or maximize the objective function while satisfying the constraints. Components of an optimization problem are as follows:

(a) Objective Function: An objective function, which we want to minimize or maximize.
(b) Design Variables: A set of unknowns or variables, which affect the value of the objective function.
(c) Constraints: A set of constraints that allow the unknowns to take on certain values, but exclude others.

Furthermore, an optimization problem needs to define following elements:

(a) Model: Modeling is the process of identifying objective function, variables and constraints. The goal of model is “insight” not the numbers. A good mathematical model of the optimization problem is needed.
(b) Algorithm: Typically, an interesting model is too complicated to be able to solve in with paper and pencil. An effective and reliable numerical algorithm is needed to solve the problem. There is no universal optimization algorithm. Algorithm should have robustness (good performance for a wide class of problems); efficiency (not too much computer time) and accuracy (can identify the error).

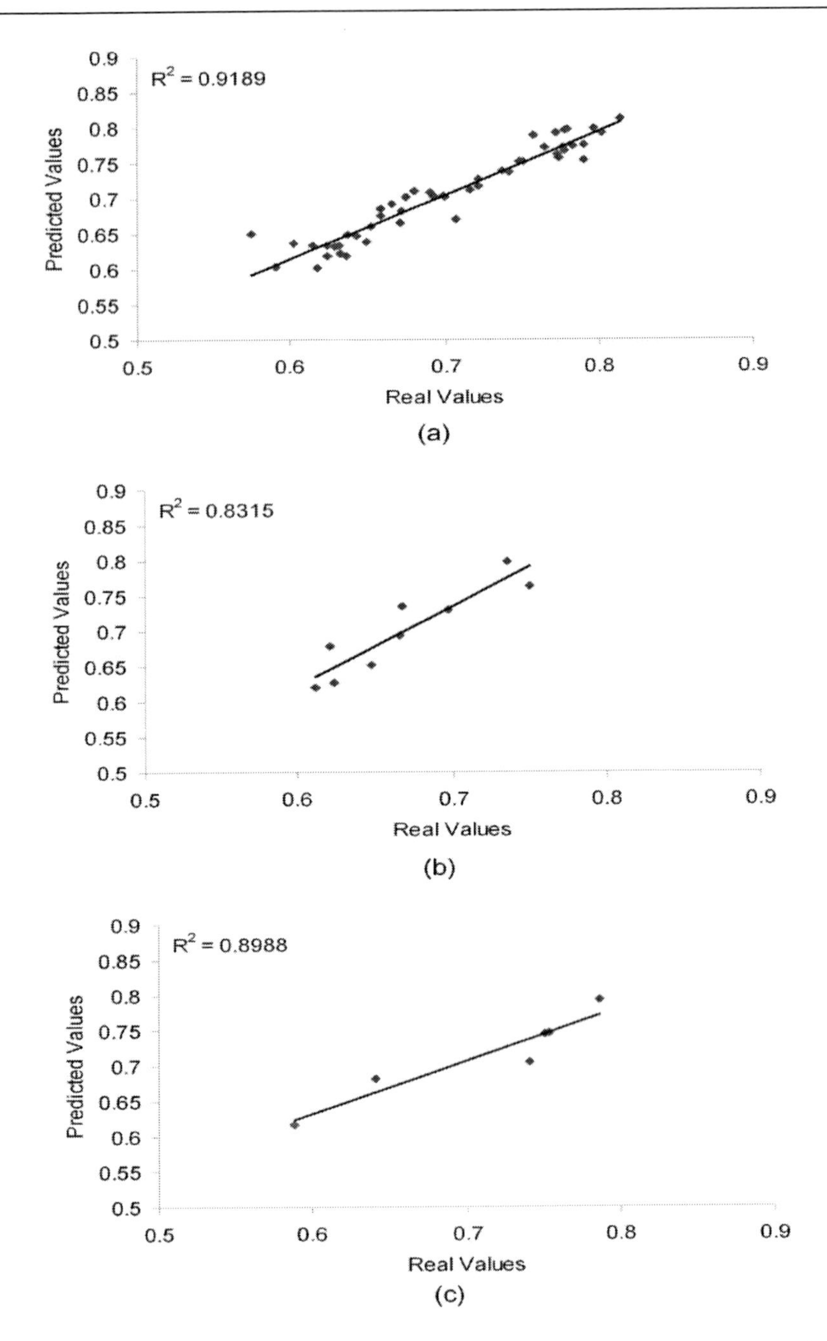

Figure 9. The determination coefficient of train, validation and test sets. (a) train (b)validation (c) test.

This neural network model was introduced to the optimization algorithms as the objective function.

In our system, optimization problem can be expressed inthe following:

Find: Power, Speed, QSF and PW
Maximize: Gloss (Power, Speed, QSF and PW)

Subject to the following variable bounds:

$-1 \leq$ Power, Speed, QSF and PW ≤ 1

These variables bounds are in the range of valid neural network modeling boundary. Since we normalized the parametersto the boundary [-1, 1] and ANNs were trained with thesenormalized parameters, the only parameters that belong to thisboundary could have valid ANN output [13].

Also, the elements of our optimization problem are introduced in the following:

Model: The best obtained evaluator from the previous section, the CFBP network with 4-9-8-1 topology and TANSIG-TANSIG-LOGSIG threshold functions.

Algorithms: Due to the complex nature of under study physical system, AI based algorithms such as GA and PSO were utilized.

5.2. Optimization Using GA

The deployment of the optimal solution search requires thetuning of some features related with the GA, for examplepopulation size, selection method and crossover functions, mutation rate, migration, etc,as mentioned in the Section 2.3. Although somegeneral guidelines about such selections exist in the relevantliteratures[45-46], but optimal setting is strongly related to the design problem under consideration and can be obtained through thecombination of the designer's experience and experimentation. On the other hand, these trial and errors need many efforts and experiments and hence are time consuming.

To conquer this problem, a new method is developed. In this method, the optimum number of population size is found and then other parameters are tuned base on the specific runs.

Amongst different parameters of GA, population size has the most important effect on the GA effectiveness. Small population size may be leads to trap into the local minima, while a high population size value causes excessive increase of each run's time. Moreover, as mentioned earlier, the trial and error procedure is a time consuming method. Therefore, the optimum selection of population size is necessary. An effective population size number, not only decreases the local minima risk, but it also controls excess running time. In this proposed method, firstly, the optimum value of population size will be determined and then, base on the trial and error, other parameters will be tuned. Following steps, illustrates optimum determination of population size:

i) Starting the optimization with a small population size and recording its result.
ii) Duplicating the population size number, redoing the optimization with previous settings and recording the result.
iii) Repeating the step ii until high growth of running time or small changes in results (compared to the previous results).
iv) Now, results should be plotted on a graph. Then a quadratic or cubic (depending on the number of nodes) type of regression model should be fitted to the results. Afterwards, a straight line should be drowning from the first point to the last one. Then a tangent line, which is parallel to the straight line, should be drowning to the

fitted line. The place which this tangent line touches the curve, indicates the optimum population size value.

Now the process of trial and errors could be performed under reliable population size number.

Utilization of this method and its final result, are illustrated in table 10 and figure 10. The optimal GA parameters setting chosen in the present study isdesigned using MATLAB GA toolbox[47].

It is seen in the table 10 that increasing the population size from 160 to 320, does not improve the optimization result; consequently, the process of duplicating the population size was halted. Additionally, by implementing the step iv, it is seen that the place which this tangent line touches the curve is related to the number of 148. Therefore, this population size number was fixed and then the process of trial and error was performed to find the best optimization result. The best result using this procedure is shown in the table 11.

Table 10.Results of steps i, ii and iii until satisfying the halt condition

Population Size	Optimization Result
5	921
10	925
20	930
40	932
80	936.51
160	936.71
320	936.68

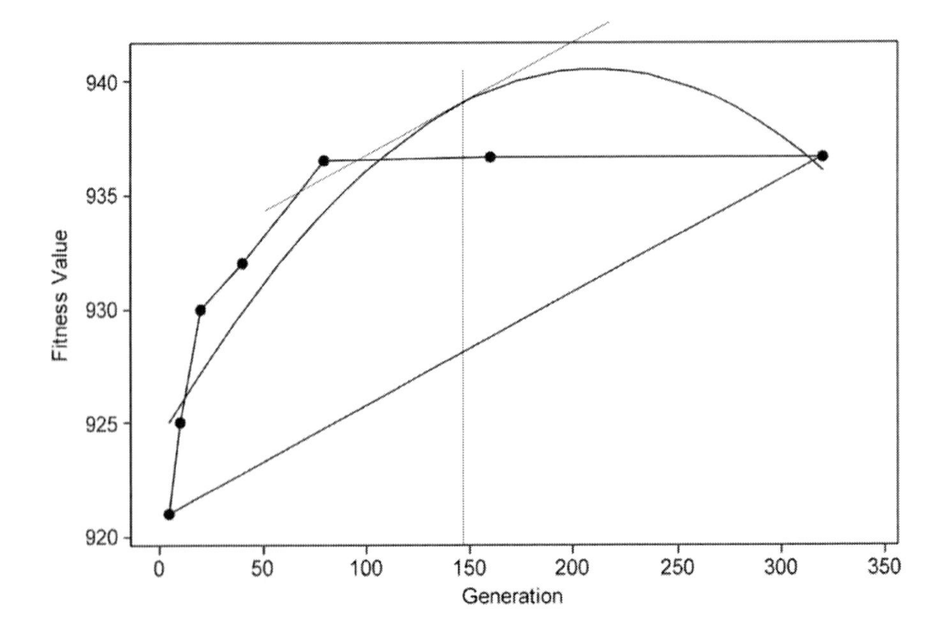

Figure 10. Finding the optimum value for the number of population size.

5.3. Optimization Using PSO

PSO has been applied successfully to a wide variety of search and optimization problems and it has been proven a powerful competitor to other evolutionary algorithms such as genetic algorithms [25]. Although simple algorithm, easy to implement and few parameters to adjust are the main features of PSO algorithm, but there are still a number of related issues concerning PSO parameter selection such as controlling velocities (determining the best value for V_{max}), swarm size, neighborhood size and robust settings for c_1 and c_2.

There are some strategies and few empirical rules can be found in the manuscripts [48-50]which could be worked out to guide the effective choice of parameters.These rules are so approximate and for a given problem, we are faced with thestrong possibility of searching at length before finding a "good" set of parameters.The good news, nevertheless, is that PSO is very robust, in the sense that broadvariations in the parameters do not prevent convergence, even if, of course, it can be more or less rapid [48]. Using the empirical guidelines and a written code in the MATLAB [51] the optimum parameters of system were derived that are demonstrated in table 11.

5.4. Results and Discussion

In this paper, genetic algorithm and particle swarm optimization methods were used to determine asetting of laser marking parameters that causes a high gloss laser marked gold. The best result of each method is shown in the table 11.

Table 11. Optimization results of GA and PSO

Method	Optimum parameters				Optimization Result (calculated gloss)
	Power percent (%)	Speed (mm/s)	QSF (Hz)	PW (μm)	
GA	1	1	-0.27	0.15	937
PSO	0.96	0.96	-0.38	0.41	908

From this table,one can observe that the gloss value increases to 937 and 908 after the optimization by GA and PSO, respectively. In the other words, the calculated gloss value using GA and PSO increases by 15% and 11% respectively as compared with the non-optimized (experiment) result (812). Figure11 plots the best individual, which shows the vector entries of the individualwith the best fitness function value.

6. Methodology Validation

In order to evaluate the accuracy of calculations, after de-normalization of GA result, a verification experiment was performed. Table 12 lists the optimized and non-optimized results. From this table, it is concluded that the gloss value increases to 937 after the optimization by GA and 905 by a validation experiment (increases by 15% and 11% compared with the non-optimized 812, which the non-optimized result is the maximum glossvalue observed in the experiments). In addition, we can see that the optimal value of QSFis less than the nonoptimized result, but optimal values of power percent, speed and PW are greater than the non-optimized result.

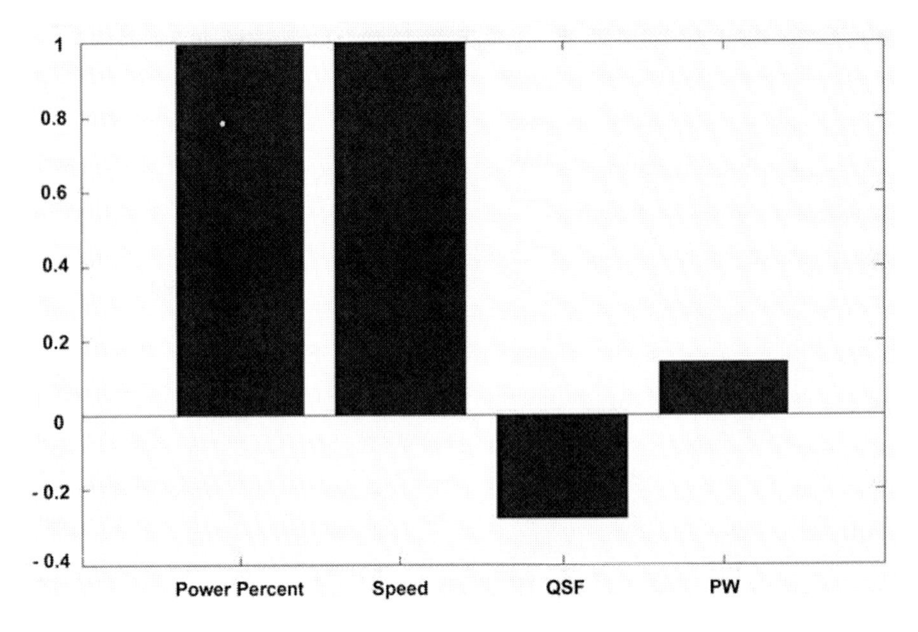

Figure 11. Vector entries of the optimum value.

Since GA optimization result is better than PSO output, the optimum calculated parameters of GA result were used for verifying experiment.

Table 12.Optimized and non-optimized results

Method/experiment	Parameters				Objective function (gloss)
	Power percent	Speed (mm/s)	QSF (Hz)	PW (µm)	
GA result (normalized)	1	1	-0.27	0.15	937
GA result (de-normalized) or validation experiment	100% (5 watt)	200	9460	25	905
Non-optimized result (experiment)	97% (4.85 watt)	150	12000	15	812

The difference between GA result and validation experiment is relating to the different sources. First, in our work we did not consider the effect of other parameters such as surface condition, geometry changes and workpiece thickness changes, and so forth, in the modeling process. Second, the errors in the gathering the experimental samples because of inaccuracy of measurement tool resulted in decreasing of modeling process accuracy which affects the optimization process.

Conclusion

In this paper, usage of artificial intelligence approaches for modeling and optimization of the effect of laser marking parameters on the gloss of the laser marked gold have been presented. After preparation of suitable training data base on the DOE approach, modeling procedure werestarted by utilization of ANFIS and ANNs. ANFIS modeling was initiated by appropriate selection of input variables for training process. An expanded method of [15]was utilized for this part and various kinds of membership functions with equal/different number of membership functions were used for determining the proper ANFIS structure. Second approach was a new method of structure setting, base on the ANOVA. Besides, ANNs were used for non-linear mapping of the laser marking parameters effects on the gloss. From this step, a CFBP network with 4-9-8-1 topology and TANSIG-TANSIG-LOGSIG threshold functions accepted as the best evaluator. For this selection, usage of a new model validation method based on optimization result was introduced. The best evaluator was presented to the optimization algorithms, GA and PSO, to solve the optimization problem. With the maximum gloss as the objective function, the optimal parameters were calculated by application of GA and PSO. After the optimization, the calculated gloss was increasedby 15% and 11% respectively, as compared with the non-optimized result. Moreover, a new method for effective determination of population size of GA was proposed. Finally, validation experiment revealed that the final gloss of laser marked gold increased by 11% as compared with the non-optimized operational parameters.

Appendix A. Comparisonof Someof ANFIS Modeling and ANN ModelingResultsbefore and afterCleaning the Data

Table A.1. Comparison of some of ANFIS modeling results before and after cleaning the data for various kinds of MFs with equal number of MFs

MFs Type		MFs Number	Train R^2	Train RMSE	Validation R^2	Validation RMSE	Test R^2	Test RMSE
gbell	bc*	2 2 2 2	0.6944	0.0737	0.0298	0.188	0.1281	0.0509
	ac*		0.8746	0.0178	0.0321	0.0966	0.1964	0.0682
gaussmf	bc	2 2 2 2	0.7018	0.0728	0.0357	0.1815	0.0329	0.1574
	ac		0.8735	0.018	0.0322	0.0957	0.1916	0.0683

*bc and ac stands for before cleaning and after cleaning respectively.

Table A.2. Comparison of some of ANN modeling results before and after cleaning the data for different neurons of hidden layers and several networks at the uniform/various threshold functions for layers

Network	Training Algorithm		Threshold Function	No. of Hidden Layers Neurons	Train		Validation		Test	
					R^2	*RMSE*	R^2	RMSE	R^2	RMSE
CFBP	LM	bc	LOGSIG	3-1	0.7583	0.0627	0.5939	0.0757	0.6716	0.0862
		ac			0.7813	0.0288	0.331	0.0429	0.7768	0.0546
		bc	TANSIG	9-8	0.8916	0.0486	0.7588	0.1071	0.724	0.1436
		ac			0.992	0.0057	0.8253	0.0273	0.8043	0.0459
		bc	T-L-L	6-6	0.9921	0.012	0.7139	0.062	0.8227	0.0674
		ac			0.9316	0.0161	0.8295	0.0442	0.9433	0.1951
FFBP	LM	bc	LOGSIG	3-3	0.6467	0.0798	0.6802	0.0789	0.6128	0.1551
		ac			0.7307	0.032	0.533	0.043	0.5504	0.0945
		bc	T-L-T	4-2	0.8070	0.0599	0.5554	0.083	0.5771	0.1083
		ac			0.8014	0.0275	0.6314	0.0391	0.232	0.086
		bc	T-T-L	11-9	0.9852	0.0184	0.8023	0.0669	0.8958	0.0552
		ac			0.9524	0.014	0.8345	0.0438	0.8425	0.0513

REFERENCES

[1] Hall DR. Carbon dioxide lasers. In: WEBB C, JONES J, editors. *Handbook of Laser Technology and Applications Philadelphia,* PA, USA: IOP Publishing; 2004. p. 1601-2.

[2] Ready JF, Farson DF. LIA handbook of laser materials processing. Orlando, USA: Laser Institute of America, Magnolia Publishing; 2001.

[3] Bruton NJ. Profiling Laser Coding in the Packaging Industry. *Optics and Photonics News*. 1997;8(5):24.

[4] McKee TJ. How lasers mark. *Electrotechnology*. 1996;7(2):27-31.

[5] Obein G, Knoblauch K, Viéot F. Difference scaling of gloss: Nonlinearity, binocularity, and constancy. *Journal of Vision*. 2004;4(9):711-20.

[6] Kigle-Boeckler G. Measurement of gloss and reflection properties of surfaces. *Met. Finish.* 1995;93(5):28-31.

[7] Keyf F. Evaluation of gloss changes of two denture acrylic resin materials in four different beverages. *Dental Materials*. 2004;20(3):244-51.

[8] Kamiya A, Ovaska SJ, Roy R, Kobayashi S. Fusion of soft computing and hard computing for large-scale plants: a general model. *Applied Soft Computing*. 2005;5(3):265-79.

[9] Ku CC, Lee KY. Diagonal recurrent neural networks for dynamic systems control. Neural Networks, *IEEE Transactions on*. 2002;6(1):144-56.

[10] Jang JSR. Self-learning fuzzy controllers based on temporal backpropagation. Neural Networks, *IEEE Transactions on*. 2002;3(5):714-23.

[11] Jang JSR, Sun CT. Neuro-fuzzy modeling and control. *Proceedings of the IEEE*. 2002;83(3):378-406.

[12] Jang JSR, Sun CT, Mizutani E. Neuro-fuzzy and soft computing: a computational approach to learning and machine intelligence. Upper Saddle River, NJ, USA: Prentice-Hall; 2002.

[13] Deemuth H, Beale M, Hagan M. *Neural network toolbox for use with MATLAB,* User's Guide. 2007.

[14] Kasabov NK. *Foundations of neural networks, fuzzy systems, and knowledge engineering*. Massachusetts, USA: The MIT Press; 1996.

[15] Jang JSR, editor. Input selection for ANFIS learning. Fuzzy Systems, 1996, *Proceedings of the Fifth IEEE International Conference on 1996*; New Orleans, LA , USA IEEE.

[16] Sivanandam SN, Sumathi S, Deepa SN. *Introduction to fuzzy logic using MATLAB*. Berlin, Germany: Springer Verlag; 2007.

[17] Sugeno M, Yasukawa T. A fuzzy-logic-based approach to qualitative modeling. *IEEE Transactions on fuzzy systems*. 1993;1(1):7-31.

[18] Takagi T, Sugeno M. Fuzzy identification of systems and its applications to modeling and control. *IEEE Trans Syst, Man Cybern*. 1985;2:116-32.

[19] Jang JSR. ANFIS: Adaptive-network-based fuzzy inference system. *IEEE Transactions on systems, man and cybernetics*. 1993;23(3):665-85.

[20] Mamdani EH, Assilian S. An experiment in linguistic synthesis with a fuzzy logic controller. *International Journal of Man-Machine Studies*. 1975;7(1):1-13.

[21] Hagan MT, Menhaj MB. Training feedforward networks with the Marquardt algorithm. Neural Networks, *IEEE Transactions on*. 2002;5(6):989-93.

[22] Kalyanmoy D. Optimization for engineering design: *Algorithms and examples*. New Delhi, India: Prentice-Hall of India; 2005.

[23] De Jong KA. *Analysis of the behavior of a class of genetic adaptive systems. Michigan*: University of Michigan; 1975.

[24] Goldberg DE, Deb K, Korb B, editors. Don't worry, be messy. *The fourth international conference on genetic algorithms;* 1991; Los Altos, CA, USA: Morgan Kaufmann Publishers.

[25] Kennedy J, Eberhart RC, editors. Particle swarm optimization. Neural Networks, Proceedings, *IEEE International Conference on* 1995 27 Nov 1995 - 01 Dec 1995 Perth, WA , Australia Perth, Australia.

[26] Abido AA, editor. Particle swarm optimization for multimachine power system stabilizer design. *Proc. Power Eng. Soc. Summer Meeting* 2002 15 Jul 2001 - 19 Jul 2001 Vancouver, BC , Canada: IEEE.

[27] Li Y, Chen X. Mobile robot navigation using particle swarm optimization and adaptive NN. *Advances in Natural Computation*. 2005:628-31.

[28] Messerschmidt L, Engelbrecht AP. Learning to play games using a PSO-based competitive learning approach. *Evolutionary Computation, IEEE Transactions on*. 2004;8(3):280-8.

[29] Eberhart RC, Kennedy J, editors. A new optimizer using particle swarm theory. Micro Machine and Human Science, MHS '95, *Proceedings of the Sixth International Symposium on2002*; Nagoya , Japan IEEE.

[30] ASTM. D523-89 *Standard Test Method for Specular Gloss*. 1989.

[31] DIN. DIN 67 530, Reflektometer als Hilfsmittel zur Glanzbeurteilung an ebenen Anstrich-und Kunststoffoberflachen. 1982.

[32] ISO. International Standard *2813 Paints and Varnishes-Measurement of Specular Gloss of Nonmetallic Paint Films at 20°, 60° and 85°*. 1978.

[33] JIS. Japanese Industrial Standard. *Z8741 Method of Measurement for Specular Glossiness*. 1983.

[34] Cox E. *Fuzzy modeling and genetic algorithms for data mining and exploration*. San Francisco, USA: Morgan Kaufmann Pub; 2005.

[35] Myatt GJ. *Making sense of data: a practical guide to exploratory data analysis and data mining*. Hoboken, NJ, USA: John Wiley and Sons Inc; 2007.

[36] Kermani BG. Modeling oligonucleotide probes for SNP genotyping assays using an adaptive neuro-fuzzy inference system. *Sensors and Actuators B: Chemical*. 2007;121(2):462-8.

[37] Chiu SL. Selecting input variables for fuzzy models. *Journal of Intelligent and Fuzzy Systems*. 1996;4:243-56.

[38] Adineh VR. Prediction of Operational Parameters in a CO2 Laser for Laser Cutting Using ANFIS. *Proc. National Manufacturing Engineering Conference*; Najafabad, Iran.2008.

[39] Lazi ŽR. *Design of experiments in chemical engineering: a practical gu*ide. Weinheim, Germany: Wiley-VCH Verlag GmbH and Co. KGaA; 2004.

[40] Montgomery DC. *Design and analysis of experiments*. Hoboken, NJ, USA: John Wiley and Sons Inc; 2008.

[41] Sivanandam SN, Sumathi S, Deepa SN. *Introduction to neural networks using MATLAB 6.0*. New Delhi, India: Tata McGraw-Hill; 2006.

[42] Adineh VR, Aghanajafi C, Dehghan GH, Jelvani S. Optimization of the operational parameters in a fast axial flow CW CO2 laser using artificial neural networks and genetic algorithms. *Optics and Laser Technology*. 2008;40(8):1000-7.

[43] Fernandes FAN, Lona LMF. Neural network applications in polymerization processes. *Brazilian Journal of Chemical Engineering*. 2005;22:401-18.

[44] Menhaj MB. *Computational intelligence, Volume 1: Fundamentals of Neural Networks*. Tehran, Iran: Amirkabir University Press; 2000.

[45] Goldberg DE. *Genetic algorithms in search, optimization, and machine learning*. New York, NY, USA: Addison-wesley; 1989.

[46] Goldberg DE. *The design of innovation: Lessons from and for competent genetic algorithms*. Massachusetts, USA: Kluwer Academic Publishers; 2002.

[47] Genetic algorithm and direct search toolbox 2 user's guide: The Mathworks Inc.

[48] Clerc M. *Particle swarm optimization*. London, UK: ISTE 2006.

[49] Jiang M, Luo YP, Yang SY. Stochastic convergence analysis and parameter selection of the standard particle swarm optimization algorithm. *Information Processing Letters*. 2007;102(1):8-16.

[50] Trelea IC. The particle swarm optimization algorithm: convergence analysis and parameter selection. *Information Processing Letters*. 2003;85(6):317-25.

[51] PSO Toolbox (for Matlab), http://sourceforge.net/projects/psotoolbox.

In: Artificial Intelligence
Editor: Brent M. Gordon, pp. 71-95
ISBN 978-1-61324-019-9

Chapter 3

AI APPLICATIONS TO METAL STAMPING DIE DESIGN

Shailendra Kumar*
Dept. of Mechanical Engineering, S. V. National Institute of Technology,
Surat-395007, Gujarat, India

ABSTRACT

Metal stamping die design is a tedious, time-consuming and highly experience based activity. Various artificial intelligence (AI) techniques are being used by worldwide researchers for stamping die design to reduce complexity, dependence on human expertise and time taken in design process as well as to improve design efficiency. In this chapter, various sheet metal operations, types of press tools, and AI techniques are briefly discussed. Further, a comprehensive review of applications of AI techniques to metal stamping die design is presented. The salient features of major research work published in the area of metal stamping are presented in tabular form. Thereafter, procedure for development of a knowledge-based system (KBS) for intelligent design of press tools is described at length. An intelligent system developed for quick design of progressive press tool is also presented. Finally, scope of future research work is identified.

1. INTRODUCTION

1.1. Sheet Metal Operations and Press Tools

Metal stampings are important structural components of automobiles, computers, refrigerators, type writers, kitchen utensils, electrical, electronics and tele-communication equipments. According to a survey in the US, some 100 000 metal stampings could be found in the average American home in the 1980s [1]. Sheet metal operations are economical and quick means of producing intricate, accurate, strong and durable metal stampings. Applications of these operations are increasing day by day due to their high productivity, low cost per part, improvement in material quality, minimum scrap material and energy

---

* E-mail: skbudhwar@med.svnit.ac.in

consumption. Sheet metal operations can broadly be classified into shearing and forming operations. Shearing operations are further classified as cropping, piercing, blanking, notching, trimming, shaving, parting off, lancing, and slitting. Forming operations are sub-classified as bending, coining, forming, stamping, embossing etc.

One of the important tasks in the production of metal stampings is the design of dies to suit the product features. During sheet metal operations, sheet metal is brought into the desired product shape by pressing metal strip between die block and punch. Single or multi-operation press tools (also known as dies) are used to perform sheet metal operations. Single operation dies perform only one operation with each stroke of the ram of press machine, whereas two or more than two operations can be performed by using multi-operation dies. Multi-operation dies may further be classified as progressive dies, compound dies and combination dies. In a progressive die (Figure1), work pieces are advanced from one station to another. At each station, one or more sheet metal operations are performed on the metal strip. The result is a finished component at the last station of die with every stroke of press. Progressive dies are costlier as compared to single-operation dies, but the saving in total handling costs by progressive fabrication in mass production as compared with a series of single operations may be great enough to justify the cost of progressive dies.

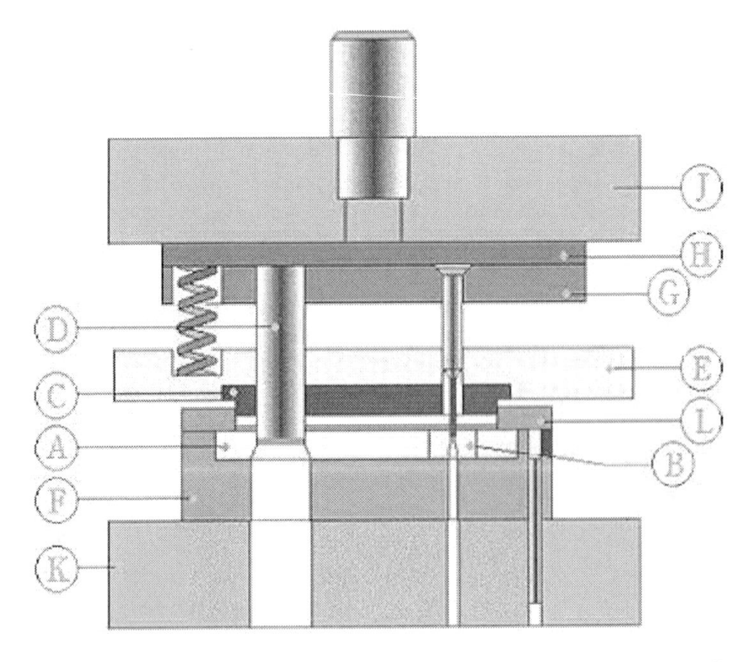

A- Die Block, B – Die insert, C – Stripper plate, D – Punches, E – Backup plate, F – Die support, G – Punch plate, H – Punch backing plate, J – Top plate, K – Bottom plate, and L – Die gage.

Figure 1. A typical two station progressive die.

1.2. Design of Press Tools

Design of press tools (or metal stamping dies) is a complex and highly specialized procedure and typically it takes 20% of the lead time from the concept design to the final stamping manufacture. The diverse nature of products produced by press tools demands a high level of knowledge on part of the die designer that can only be achieved through years of practical experience. Checking the design features of sheet metal parts, design of strip-layout / process planning, selection of type and size of die components, selection of materials for die components; and modeling of die components and die assembly are major activities for designing a press tool. The traditional methods of carrying out these tasks require expertise and are largely manual and therefore tedious, time consuming and error-prone. Also the knowledge gained by die design experts after long years of experience is often not available to others even within the same company. It creates a vacuum whenever the expert retires or leaves the company. Modern CAD/CAM technology, new ideas in design and construction of press tools, coupled with increased speed and rigidity of the presses have contributed towards the continual use of metal stamping production processes to manufacture increasingly more sophisticated products. However, these developments demand greater skills of the designers. In rapidly developing countries, the problem is further compounded by two factors –

1. Young technical educated people do not have patience to undergo long periods of apprentice training to acquire the skills to be a good die designer. Hence, there are fewer people joining the trade of die design. Stamping industries are facing acute shortage of skilled die designers. Further, the mobility of experienced die designers in stamping industries has caused much inconvenience to sheet metal industries all over the world.
2. Owing to the rapid changes in consumer taste, the products have very short life cycles. In other words, there is a rapid demand for more sophisticated metal stampings. The die designers are under constant pressure to exploit the latest design and manufacturing technology in order to meet the market demand.
3. Currently, some sheet metal industries are using traditional CAD/CAM systems to design and manufacture press tools. These CAD/CAM systems help to improve productivity of the designers by providing interactive graphical aids and a common geometrical database for them to model the components, produce the drawings and generate the NC codes for fabricating the components. However, there is a limit to the amount of productivity gain that these CAD/CAM systems can offer to a company.

To overcome the above problems, there is need to develop an intelligent system for die design using some suitable artificial intelligence (AI) technique(s) to assist die designers and process planners working in sheet metal industries, especially in small and medium sized stamping industries.

1.3. Artificial Intelligence (AI)

Artificial Intelligence (AI) is the branch of computer science dealing with the design of computer systems that exhibit characteristics associated with intelligence in human behavior, including reasoning, learning, self-improvement, goal-seeking, self-maintenance, problem solving and adaptability [2]. In a broad sense, it is the technique concerned with the development and application of computational tools that mimic or are inspired by natural intelligence to execute tasks with performances similar to or higher than those of natural systems (humans). The first application of AI in engineering in general and in manufacturing in particular was reported in the late 1980s. The importance of AI in manufacturing has clearly increased over the last 30 years. It has gradually progressed by receiving attention from the production community. AI is now used in many different areas in manufacturing engineering such as product design, production processes, operations, and fault detection etc. For the sake of simplicity, these techniques are grouped into the following categories:

(i) Knowledge Based Systems (KBS) / Expert Systems (ES),
(ii) Neural Networks (NN),
(iii) Fuzzy Logic (FL),
(iv) Multi Agents (MA), and
(v) Others such as Genetic Algorithms (GA), Simulated Annealing (SA), etc.

The use of AI techniques in sheet metal process planning and die design was started in late 1980 and early 1990s. Various AI techniques used in metal stamping die design are shown in Figure2. A brief description of these techniques is given as under.

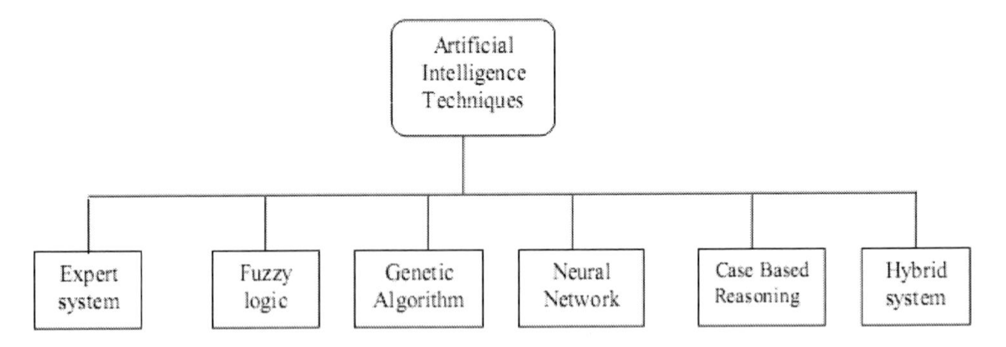

Figure 2. Artificial Intelligence (AI) techniques.

Knowledge Based System (KBS) /Expert System (ES)

Knowledge based systems (KBSs) / Expert systems (ESs) are computer programs embodying knowledge about a narrow domain for solving problems related to that domain [3]. A KBS or ES usually comprises three main elements - a knowledge base, an inference mechanism and a user interface. The knowledge base contains domain knowledge, which may be expressed as any combination of IF-THEN rules, factual statements, frames, objects, procedures and cases. The inference mechanism allows manipulating the stored knowledge for solving problems. Knowledge manipulation methods include the use of inheritance and constraints (in a frame-based or object-oriented KBS), the retrieval and adaptation of case example (in a case-based KBS) and the application of inference rules (in a rule-based KBS) according to some control procedure (forward chaining/data-driven and backward chaining/goal-driven) and search strategy procedure (depth first and breadth first).

Neural Network (NN)

Neural networks are massively parallel inter-connected networks of simple (usually adaptive) elements and their hierarchical organizations which are intended to interact with objects of the real world in the same way as biological nervous systems do. Neural networks have been inspired both by biological nervous systems and mathematical theories of learning. They belong to a family of models that are based on a learning-by-example paradigm, in which problem solving knowledge is automatically generated according to actual examples presented to the network. Neural networks have several advantages that are desired in manufacturing practice. The basic components of a neural network are nodes (neurons) and weights (connections). The nodes are usually characterized by an internal threshold and by the type of their activation function. The adjustable weights correspond to biological synapses. A positive weight represents an excitatory connection. A negative weight represents an inhibitory connection. The weighted inputs to a neuron are accumulated and then passed to an activation function, which determines the neuron's response. Neural networks are characterized by their learning ability and parallel-distributed structure.

Case Based Reasoning (CBR)

Case-based reasoning technique was devised to index, retrieves and modifies designs, which have been carried out previously to produce new designs. There are several advantages of using a CBR approach in die design. Firstly, the CBR approach reasons out design solutions quickly by searching and reusing past design cases, thus avoiding the need to design everything from scratch. Secondly, the CBR approach possesses the ability to learn from past mistakes by remembering previous experiences.

Hybrid System

Each of the above-mentioned AI based approaches used in metal stamping die design has some advantages and disadvantages. One approach to deal with complex metal stamping die design problems is to integrate the use of several AI techniques in order to combine their different strengths and overcome a single technology's weakness to generate hybrid solutions. This hybrid system provides a cooperative decision making environment and facilitates a hybrid knowledge representation scheme, including procedures, production rules, object-oriented, graph-based and case based representations neural network fuzzy logic etc. Such a type of approach can speed up die design process.

2. Review of Applications of AI Techniques to Metal Stamping Die Design

2.1. Manufacturability Evaluation of Sheet Metal Parts

In order to develop better concurrency between design and manufacturing of metal stamping dies, a stampability assessment or evaluation of sheet metal parts is necessary. It is estimated that decisions made at the part design stage determine 70%–80% of the manufacturing productivity [4]. Therefore, as a first step in the planning for manufacture of a sheet metal part, it is useful to check its internal as well as external features for assessing its manufacturability on stamping die. Over the years, the industrial practices of checking of the internal and external features of sheet metal parts have not changed significantly. Traditional methods involve calculations and decisions, which have to be made on the basis of experience and practice without the computer aids. Later on, some AI techniques were developed by researchers. But die designers realized the applications of AI techniques around 1980s. The assessment of manufacturability aspect is done by using various approaches such as neural network (NN), fuzzy logic (FL), agent-based system (ABS), rule-based system (RBS), analytical hierarchy process (AHP), object oriented techniques (OOT), and case-based reasoning (CBR).

The Technology Check module of the Computer Aided Die Design System (CADDS) proposed by Prasad and Somasundaram [5] is capable to assess the feasibility of the given sheet-metal blank for blanking process. It checks whether the given blank profile can be easily produced by sheet metal stamping through a single operation die. A checking algorithm has been implemented to prove the acceptability of the blank profile. It also suggests the designer about the possible changes in blank profile for easier manufacturing. But the major limitation of this system is that it is limited to check the manufacturability of sheet metal parts only for single operation blanking and piercing dies. Lazaro et al. [6] developed a KBS for identifying design rule violations to improve part manufacturability. Shpitalni and Saddan [7] addressed the problem of automatic tool selection and bending sequence determination using graph search heuristics. Meerkamm [8] developed a system to detect design violations concerning manufacturability of sheet metal parts. Lee et al. [9] developed an assessment system consisting of a knowledge-based geometric analysis module, a finite element module and a formability analysis module. The geometric analysis module uses geometric reasoning and feature recognition with a syntactic approach to extract high-level geometric entity information from vertices in two-dimensional forming. The empirical rules for stamping die design are represented as frames in the knowledge base. Mantripragada et al. [10] developed a feature based design system, which acts like an interactive design tool and can be used to alert designers for potential production problems, defects and failures. Yeh et al. [11] developed a rule-based design advisor for sheet metal parts. This system also suggests alternatives redesign solution and estimates cost. Radhakrishnan et al. [12] presented the advisory design rule checker system, integrated into ProMod-S using medial axis transformation algorithm to check the number of features for complicated sheet metal parts. Gupta et al. [13] described a process planning system for robotic sheet metal bending press. The system automatically determines bending sequences, selection of punches and dies and manufacturing costs etc. and gives feedback to improve the plan on operation-by-operation basis.

Xie et al. [14] developed a compound cutting and punching production method for small and medium size sheet metal industries. The system uses concurrent and global design and manufacturing environments by integrated data integration platform based on Pro/INTRALINK and STEP, and a knowledge-based real time CAPP (RTCAPP) system into existing CAD system. Tang et al. [15] proposed an intelligent feature based design for stampability evaluation of a sheet metal part for quick check of potential problems in stamping process and stamping die. They integrated design evaluation and cost estimation in a single system. Ramana and Rao [16] presented a system for automated manufacturability evaluation. The system described design evaluation, process planning, data and knowledge modeling for shearing and bending operations. Giannakakis et al. [17] presented an expert system for process planning and die design. This system includes initial calculations, process planning, die and press selection, and tools selection modules for cutting and piercing operations of progressive die. It is coded into CLIPS expert system development environment. Shukor and Axinte [18] discussed key issues and technologies related to the construction of manufacturability analysis systems (MAS).

Salient features of major research work in the area of manufacturability evaluation of sheet metal parts are summarized in Table 1.

Table 1. Salient features of major research work in the area of manufacturability evaluation of sheet metal parts

Ref. No.	Researchers	Input Mechanism	AI Technique(s) used	Remarks
[5]	Prasad and Somasudaram (1992)	User-system interactions	Rule-based system	For blanking and piercing parts only.
[6]	Lazaro et al. (1993)	User-system interactions	Used Rule based and object oriented tech.	Identified design rule violations and suggested alternatives redesign solutions to improve manufacturability.
[9]	Lee et al. (1995)	CAD model	Geometric reasoning and feature recognition	An assessment system consisting of a geometric analysis module, a finite element method module and a formability analysis module.
[10]	Mantripragada et al. (1996)	CAD model	Knowledge based system	Formability analysis of sheet metal parts and gives redesign solutions. Limited to simple and specific part for checking manufacturability of sheet metal parts
[11]	Yeh et al. (1996)	Collection of manuf. information and user-system interactions	Rule based expert system	Design advisory system for sheet metal parts that suggests alternatives redesign solutions and estimates cost.
[12]	Radhakrishnan et al. (1996)	CAD model	Geometric reasoning	Rule checker and design advisory system for holes, slots, and other features of sheet metal part.
[14]	Xie et al. (2001)	CAD model	Knowledge based system	Automatic tool selection, manufacturing sequencing and cost estimation module for small- and medium-sized job shops
[15]	Tang et al. (2001)	CAD model	Feature based and rule based expert system	Stampability evaluation system for sheet metal parts
[16]	Ramana and Rao (2005)	CAD model	Rule based expert system	Generation of evaluation and process plan for sheet metal parts in mass production
[17]	Giannakakis and Vosniakos (2008)	User-system interactions	Knowledge based system	Limited to cutting and piercing operations only

2.2. Process Planning and Metal Stamping Die Design

Research and development of die design automation systems was given a new dimension in the late 1980s and early 1990s, when the applications of AI techniques in engineering design started to take off. One area of the research that has attracted a considerable amount of researchers is development of knowledge-based die design and process planning systems.

Lin et al. [19] developed PC based expert system using FORTRAN, Micro Expert and AutoCAD for design of die-set of blanking and simple drawing die. Sitaraman et al. [20] proposed a KBS for process sequence design in axisymmentric sheet metal forming. Lee et al. [21-22] developed IKOOPP, a knowledge-based process planning system for the manufacture of progressive die plates. IKOOPP is able to automatically recognize the machining features from 3D die plate modeled in Auto-trol and proceed to automatically plan the set up sequences, select the required machine tools, cutting tools, heat treatment, fixturing elements and sequence of operations. Duffy and Sun [23] from University of Massachusetts described a proof-of-concept system for progressive die design for simple hinge part. The system is implemented using knowledge collected from manufacturability data, industry experts and standard die components. The system generates flat pattern geometry and develops a strip layout automatically. Li et al. [24] developed a knowledge based CAD/CAM package for progressive dies for small sheet metal parts. Prasad and Somasundaram [5] developed a computer aided die design system (CADDS). The system is capable to generate strip-layout automatically, conduct design checks for various die components, and generate the assembly views and bill of materials for blanking die. This system is developed by interfacing AutoCAD with AutoLISP. Nee and Foong [25] reviewed the techniques employed in punch design automation for progressive dies and made an attempt to link the programs together to form a useful package for the design of progressive dies. The program incorporates knowledge-based tools, heuristic rules, and number of expert design rules. Researchers at University of Liverpool [26-29] have also applied efforts for the development of expert systems for progressive piercing and blanking die design. Their research concentrated on shape-coding and recognition techniques for the decomposing of the bridge scrap into smaller shapes. However, their techniques are limited to straight-edge workpieces only. They also developed a UNIX-based expert system for the planning and the design of progressive piercing and blanking dies. The system was developed by integrating Auto-CAD with Kappa and some C programs. They stressed for the development of knowledge-based system for selection of die components. Tisza [30] developed a Knowledge-based expert system using principles of group technology for multi-stage forming process for feature recognition, material selection, blank determination, optimum sequencing of operations, tool and machine selection. Cheok et al. [31, 32] from National University of Singapore reported to develop an intelligent progressive die (IPD) design system. They used various AI techniques such as feature-based modeling, rule-based approach and spatial reasoning to work-piece shape representation, shape recognition and decomposition, and die component representation for die design automation. Esche et al. [33] explored KBS to design intermediate tooling conditions and determined minimum drawing steps of axisymmetric deep drawing problems and two-dimensional forming problems. Ong et al. [34] developed a system using an algorithm for automatic generation of bending sequence for progressive die. It consists of several modules, namely feature recognition, strip layout and die design. Park [35] developed an expert system for electronic gun grid parts. This system is coded in C under UNIX platform and CIS customer language of the EXCESS CAD/CAM system. Choudhary and Allada [36] developed an integrated PC based CAD/CAM system for design of precision punches and die for small scale manufacturer. Singh and Sekhon [37] developed a low cost expert system for small and medium sized enterprises to assist sheet metal planner in making an intelligent selection of press machine from alternative machines available. The system is coded using AutoLISP programming language. Pilani et al. [38] proposed a neural network and Knowledge-based, hybrid intelligent system approach to generate an optimal die face for forming dies. Caiyuan et al. [39] reported to develop a knowledge-based CAD/CAM package labeled as HPRODIE for progressive dies for small-size metal stampings using feature mapping, rule based reasoning and case-based reasoning techniques.

Choi and Kim [40] developed a compact and practical CAD/CAM system for the blanking or piercing of irregular shaped-sheet metal products for progressive working. Shi et al. [41] developed a process planning system for an auto panel. Tor et al. [42] used CBR and graph-based approach for progressive die design and process planning. It combines the flexibility of blackboard architecture with case-based reasoning. This system is capable of managing heterogeneous knowledge sources. Vosniakos et al. [43] presented logic programming paradigm for checking part design, configuration of press tools and planning process for U shape bend part. This system is coded using PROLOG language. Dequan et al. [44] presented a comprehensive review of KBS in stamping planning. They presented a framework of CAD system that carries out automated process planning for piercing operation of precision work at a high speed. Chan et al. [45] developed an integrated system using FEM simulation and artificial neural network (ANN) to approximate the functions of design parameters and evaluate the performance of die designs before die tryout. Chi [46] proposed the fuzzy classification and rough set approach for mining the die design knowledge from various resources. These mined rules guide the designer in stamping die design. Zhibing et al. [47] developed a multi-step unfolding method (MSUM) for blank design and formability prediction of complicated progressive die stamping part. Chu et al. [48] used graph theoretic algorithm for automatic operation sequencing for progressive die design. The algorithm is implemented in C++ and is fully integrated with SolidWorks CAD system. Ghatrehnaby and Arezoo [49] developed algorithm for an automated nesting and piloting system for progressive dies. Their work is concentrated on geometrical optimization of nesting and piloting in CAD system. Tsai et al. [50] developed a system to automate process planning and die design in automotive panel production using knowledge based engineering methodology.

A summary of major research work in the area of applications of AI techniques to process planning and metal stamping die design is presented in Table 2.

Table 2. Major research work in the area of AI applications to process planning and metal stamping die design

Ref. No	Researchers	System Details	Remarks
[19]	Lin et al. (1989)	Developed PC based expert system using FORTRAN, Micro Expert and AutoCAD for progressive die	Limited to simple blanking and piercing of 2D parts having simple geometrical profile
[20]	Sitaraman et al. (1991)	KBS for process sequence design in axi-symmetric sheet metal forming.	Limited to certain phases of process planning of forming die design process.
[21,22]	Lee et al. (1991, 1993)	Developed IKOOPS for process planning. Used object-oriented schema together with production rules and heuristics.	Developed for the planning of progressive die plates
[23]	Duffy and Sun (1992)	Used concept of KBS to model die components and assemblies for hinge part. Implemented in ICAD design language.	Developed for design of door hinges. Requires an experienced die designer to operate the system.
[24]	Li et al. (1992)	Rule based KBS for small size electronic parts using graph theory.	Limited to specific application

Table 2. (Continued)

Ref. No	Researchers	System Details	Remarks
[5]	Prasad and Somasundaram (1992)	Developed KBS using AutoCAD with FORTRAN 77 and AutoLISP for automatic generation of strip-layout, die-layout, and bill of materials.	Developed for blanking operations and for simple parts.
[25]	Nee and Foong (1992)	Punch shape design automation for progressive dies using heuristic rules	Limited to punch shape design and not capable to check the design features of sheet metal parts
[26-29]	Ismail et al. (1993, 95, 96)	Rule-based object oriented and feature-based KBS for punch shape selection, number of staging and size of die components	Developed for straight edge sheet metal parts of blanking and piercing operations
[30]	Tisza (1995)	KBS for generation of process sequences and designing of tools for deep drawing of axi-symmetric and rectangular part.	Required inputs from skilled die designer to operate the system.
[31]	Cheok and Foong (1996)	Linking of a CAD system, a KBS and library of numerical routines for design of progressive die	Not capable of checking design features of sheet metal parts
[33]	Esche et al. (1996)	Rule-based KBS for generation of forming process outlines for round shaped parts.	Inputs from die-designer required and an experienced die designer is required to operate the system
[34]	Ong et al. (1997)	Described a methodology to determine the optimal set-up and bending sequences for the brake forming of sheet metal components	Limited to specific application
[32]	Cheok and Nee (1998)	Intelligent KBS for progressive die design	Unable to check design features of sheet metal parts. Also the designer is expected to generate the assembly views and bill of materials interactively.
[35]	Park (1999)	KBS approach to develop CAD/CAM system for deep drawing die using object-oriented techniques	Developed for blanking die set and simple deep drawing press
[36]	Choudhary and Alled (1999)	Rule-based KBS for design of precision punches and die, written in AutoCAD platform using AutoLISP programming language.	Developed for a specific applications and small parts
[37]	Singh and Sekhon (1999)	Rule-based KBS approach for optimum selection of press	Need user interactions and die design knowledge
[38]	Pilani et al. (2000)	A hybrid system using ANN and KBS approach to optimal design the die faces of forming die	Experienced die designer is required and developed for specific applications
[40]	Choi and Kim (2001)	Used KBS approach and developed by embedding algorithms in AutoCAD. AutoLISP language is used for programming.	Developed for mainly blanking or piercing operations of irregular shaped sheet metal parts
[41]	Shi et al. (2002)	KBS for processing planning for auto-panel using rule base and case base	Concentrate on specific problem
[42]	Tor et al. (2003)	Used CBR to develop KBS for design of progressive die	Need initial cases as database, also indexing and retrieving of cases from database makes design process slow
[16]	Ramana and Rao (2005)	Rule-based KBS for checking the manufacturability of sheet metal parts	Need to focus on the effect of change in design parameter on functionality of part
[43]	Vosniakos et al. (2005)	Used logic programming paradigm to assess part and tool design, as well as to suggest process plans	Limited to U shape parts, cutting and piercing operations only
[44]	Dequan et al. (2006)	KBS for Process planning for stamping die using Object-orient techniques	Limited to specific problem, need to focus on optimization of material utilization.
[45]	Chan et al. (2008)	Developed an integrated methodology using ANN and FEM to address the uncertainties in design of metal-formed part	Experienced designer is required
[47]	Zhibing et al. (2008)	Finite element model is developed using AI for blank design and formability prediction of complicated progressive die stamping part	Finite element analysis calculation consumes lot of time
[50]	Tsai et al. (2010)	CBR is integrated into ordinary process planning and die design processes to generate a hybrid KBE system for automotive panel production	Limited to specific application

2.3. Comments on Reviewed Literature

Some researchers have developed special CAD/CAM systems for metal stamping work. But most of these systems are applicable only to specialized parts or parts with relatively simple geometry. Commercially available CAD/CAM systems are providing some assistance in drafting and analysis in die design process, but human expertise is still needed to arrive at the final design. Also, the high cost associated with setting up such systems is quite often beyond the reach of small-scale sheet metal industries. Further, Majority of the developed CAD/CAM systems for progressive die cannot perform many tasks such as checking of design features of sheet metal parts, design of strip-layout, selection of press tool components and material selection of die components, which are highly dependent on the knowledge and experience of the designers. Therefore, it is most appropriate to use knowledge-based approach to establish CAD/CAM systems for the design of metal stamping dies. The advantages of using KBS approach include the utilization of knowledge of domain experts, high efficiency and flexibility. Development of such systems for press tools can also promote systematization and standardization of knowledge in sheet metal stamping.

Some researchers have used KBS approach to conserve experienced based knowledge of die design experts. But the use of these systems is very limited. They can either handle only blanking and piercing operations or parts with relatively simple geometry. Most of the systems are developed for single-operation stamping dies using production rule-based approach of AI. Very few systems are developed for design of multi-operations dies and even these are not capable to fully automate the die design process. Thus, there is a stern need to develop a KBS for intelligent design of metal stamping dies using both CAD and AI approach collectively. The system should be capable of performing major activities of design of metal stamping dies such as checking of design features of sheet metal parts, design of strip-layout / process planning, selection of die components and material selection for die components. Further, the system should be flexible enough to accommodate any future modifications on the advancement in technology. It should be interactive and have low cost, which can be easily affordable by small and medium scale sheet metal industries. As the process of metal stamping die design comprises of many activities, the whole KBS can be structured into various sub-systems, modules and sub-modules. All system modules should be user interactive and designed to be loaded in the prompt area of AutoCAD.

To construct the system modules, author has identified a proper procedure through critical study of literature available and discussion with domain experts. The identified procedure is described as under.

3. Procedure for Development of Knowledge Base System (KBS) for Design of Metal Stamping Die

The procedure for constructing KBS modules for design of metal stamping die is schematically shown in Figure 3. A brief description of each step of the procedure is given as under [51].

3.1. Knowledge Acquisition

Knowledge acquisition for each module of press tool design is essentially a collection of bits and pieces of published or unpublished, analytical or empirical knowledge from a variety of sources including experienced progressive die designers, shop floor engineers, and handbooks, monographs, research journals, catalogues and industrial brochures. A brief description of these sources is given as under.

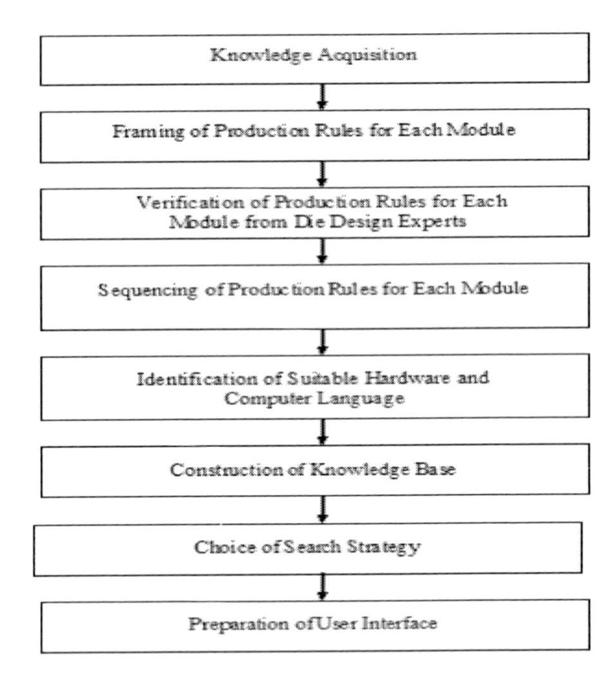

Figure3. Procedure for development of KBS for design of metal stamping die.

Literature Reviews

Review of the literature including die design handbooks and published research papers on die design and CAD provides detailed, academically fundamental information and latest research progress in these areas. Although the information obtained from the die design handbooks is not always the same as what is currently being practiced, literature survey is an easy and inexpensive mode of knowledge acquisition.

Die Design Experts

The major bottleneck in the development of KBS for die design is the process by which the knowledge of die design expert(s) is extracted and work is done in close coordination with expert(s) and serves for the adequate modeling of expertise and human inference capabilities. The process of knowledge acquisition from experienced die designers involves presenting a few typical problems to the expert(s) and letting the expert(s) talk through the solution. During the verbal analysis, the expert(s) would be questioned / interrupted to explain why a particular decision was reached. Knowledge may be acquired from domain experts by holding discussions on typical problems pertaining to the major activities of press tool design and letting the experts talk about the approach, formulae and thumb-rules relied upon by them. This is to identify the parameters influencing particular decision. However for acquiring the maximum possible knowledge in die design domain they should be provided a list of situation based questions. Experts should be requested to reply the quarries in form of various possible situation of 'IF-Then' kind of that specific domain. They should also be observed while working. Domain experts may face difficulties to represent their knowledge in form of facts, rules etc. They prefer to represent their knowledge through examples.

Industrial Visits

This knowledge, not always quantifiable, is useful in getting a better 'feel' for the problem of press tool design and in understanding the common terminology used on the shop floor of stamping industries, and thus facilitates constructing a practical KBS that would cater the needs of stamping industries and also be equipped with a friendly user interface. A KBS with such a front-end is expected to have a relatively easy learning curve and therefore, is likely to be accepted more by the industrial users.

Industrial Brochures

The information obtained through industrial brochures of stamping industries is a compromise between the academically fundamental knowledge obtained through literature reviews and the practical, experience-based knowledge obtainable from industrial experts.

3.2. Framing of Production Rules

Production rule-based systems constitute the best means available today for codifying the problem solving know-how of die design experts. The syntax of a production rule is –

IF < condition >
THEN < action >

The condition of a production rule, sometimes-called LHS (Left-Hand side) contains one or more conditions, while the action portion, sometimes called RHS (Right-Hand side) contains one or more actions.

3.3. Verification of Production Rules

The production rules framed for each module must be verified from a team of die design experts and tool manufacturers by presenting them IF-condition of the production rules and then matching their recommendations with the THEN-action part of rules. The conflicting situation in production rule(s) must be tackled through discussion with the team of die design experts consulted in the framing stage of production rules. Then the necessary modifications in production rules may further be verified from some other team of die design experts.

3.4. Sequencing of Production Rules

Production rules in each module of the KBS of die design may be arranged either in an unstructured (arbitrary) or a structured manner. In the latter case, the rules tend to be simpler and briefer because they are designed to "fire" in some hierarchical manner.

3.5. Identification of Suitable Hardware and a Computer Language

Suitable hardware elements depending upon memory requirement, processing speed and needed configuration should be selected. Today, most of the KBS modules are being developed on a PC/AT because it involves low cost. Early knowledge-based systems were written in language interfaces derived from FORTRAN. Later on, LISP and PROLOG have been won wide acceptance for building knowledge-based systems. However, the user of LISP and PROLOG languages encounters difficulties when handling design problems involving graphical information. For this reason, AutoCAD and AutoLISP have found greater acceptance for the development of KBS for metal stamping die design.

3.6. Construction of Knowledge Base

Knowledge base is a part of a KBS that contains domain knowledge, which may be expressed in form of production rules of IF-THEN variety. The inference mechanism allows manipulating the stored knowledge for solving problems. Rules and knowledge base are linked together by an inference mechanism. The user-input information provides guidance to the inference engine as to what 'IF-Then' rules to fire and what process of information are needed from the knowledge base.

3.7. Choice of Search Strategy

When searching for a solution to metal stamping die design problems, two strategies called forward chaining and backward chaining are generally used. Forward chaining is a good technique when all on most paths from any one of much initial or intermediate state converges at once or a few goal states. Backward chaining is an efficient technique to use when many goal states converge on one or a few initial states.

3.8. Preparation of User Interface

KBS modules should be user interactive in nature. The purpose of user interface in the development of KBS module is twofold -to enables the user to input the essential sheet metal component data, and to displays the optimal decision choices for the user's benefit. The former is accomplished by flashing AutoCAD prompts to the user at appropriate stages during a consultation to feed data items. Messages or items of advice are likewise flashed into the computer screen whenever relevant production rules are fired.

Using the above procedure, author has developed an intelligent system for design of progressive die (labelled as INTPDIE). The system INTPDIE is described as under.

4. AN INTELLIGENT SYSTEM FOR DESIGN OF PROGRESSIVE DIE: INTPDIE

A progressive die is used worldwide for mass production of sheet metal parts. Design of progressive die is a complex and highly specialized procedure. Author has developed an intelligent system labelled as INTPDIE for automated design of progressive die. This system is implemented on PC (Pentium 4 CPU, 2.4 GHz, 256 MB of RAM) with Autodesk AutoCAD 2004. The production rules incorporated in all the modules of the proposed intelligent system are coded in AutoLISP language. The system works with input information supplied by the user coupled with knowledge stored in the knowledge base, to draw conclusions or recommendations. The developed KBS 'INTPDIE' overall comprises of more than 1200 production rules of IF-THEN variety. A sample of production rules incorporated in various modules of the system INTPDIE is given in Table 3.

Table 3. A sample of production rules included in the system INTPDIE

S. No.	IF (Condition)	THEN (Action)
1	Sheet material ='Soft Steel' or 'Brass' or 'Aluminium', and Shape of hole = round, and Minimum round hole diameter < 0.4 mm	Set the minimum diameter of round hole = 0.4 mm
2	Sheet material = 'Hard Steel', and Shape of hole = Round, and 0.5 mm ≤ minimum hole diameter ≥ 1.3 times of sheet thickness	Accept the diameter of round hole
3	Production quantity ≥ 100000; and 0.001 < Tolerance required on part ≤ 0.2; and Number of operations ≥ 2	Design Progressive die
4	0.001 < minimum accuracy required on part in mm ≤ 0.2, and Feature required on part = small cut or notch on external boundary or contour	Required operation = Notching
5	0.001 < minimum accuracy required on part in mm ≤ 0.2; and Feature required on part = hole or slot or internal contour cut	Required operation = Piercing
6	Number of holes exist on the part ≥ 2, and Shape of holes = circular, and Diameter of holes ≥ 1.0 mm, and Hole pitch ≥ 2.0 times of sheet thickness, and Distance of holes from the edge of part ≥ 2.0 times sheet thickness, and Specified tolerance on holes ≥ ± 0.05 mm, and Holes are located on opposite sides of the part	Select the two largest holes for piloting
7	The centre to centre distance between holes has tolerance range within ± 0.05 mm	Pierce these holes at the same station

Table 3. (Continued)

S. No.	IF (Condition)	THEN (Action)
8	There are complex or weak sections in the external profile of the part	Divide it into small simple shapes for the ease of punch and die manufacturing
9	Required operations = piercing (two holes), punching, parting off, and notching	Number of stations required = 5 Preferred staging - First station: Piercing, Second station: Punching, Notching and Piloting, Third station: Notching and Piloting, Fourth station: Notching and Piloting, and Fifth station: Parting off
10	Required operations on part = Piercing (Two or more holes), and Blanking, and Distance of hole from the part edge and Hole pitch < 2.0 times of sheet thickness	No. of stations required= 3 Preferred staging - First Station: Piercing Second Station: Piercing and piloting Third Station: Blanking
11	Required tonnage > 4.0; and Required tonnage ≤ 8.0, and Required production rate/min.> 50; and Required production rate/min. ≤ 1200; and Type of sheet metal operations = shearing	Select hand or mechanical or hydraulic or pneumatic press of 10 or 20 tons capacity
12	Sheet thickness ≤ 1.6 mm; and Die material = tool steel	Select thickness of die block = 28.0 mm
13	100.0 < selected length of die (parallel to die-set) in mm ≤ 175.0; and 110.0 < selected width of die (parallel to die-set) in mm ≤ 175.0; and Tolerance required on part ≤ 0.1 mm	Place die in the 4 pillar die-set with pillar diameter 25 mm and bolster dimensions in mm as – Length=280.0, Width=280.0, and Height=30.
14	Sheet material = Cu or Al or Brass or Pb or Beryllium copper; and Shear strength of sheet material > 5 Kgf/mm^2; and Shear strength of sheet material ≤ 20 Kgf/mm^2; and Type of operations = Shearing; and Production quantity ≤ 100 000	Please select an easily available material for Punch and die/inserts from the following: UHB-ARNE (54-62 HRC) (AISI O1, W.-Nr. 1.2510) 'OR' EN-31 (56-60 HRC) (AISI 52100) (IS 103 Cr 2)

4.1. Organization of the System

As the progressive die design process involves many activities, the whole system INTPDIE has been structured into various sub-systems, modules and sub-modules [49]. Organization of the developed system INTPDIE is shown in Figure 4. Module CCKBS is developed for checking design features of sheet metal parts from manufacturability point of view. This module is capable for checking part design features such as size of blank, size of holes, hole pitch, corner radius, distance of the internal features from the edge of the part, distance between two internal features, width of recesses or slots or projections, bend corner radius etc. The data supplied by the user is also stored in a file, called as COMP.DAT for use in subsequent modules.

Module SELDIE is developed to assist the process planner in selection of a suitable type of die for manufacturing of sheet metal parts. The module is designed to take required inputs from the part data file COMP.DAT. The user is also invited to enter other required inputs involving number and type of sheet metal operations through prompt area of AutoCAD. As soon as the user enters these inputs, the module imparts intelligent advice for selection of suitable type of die.

Module MAXUTL is constructed for determining the angle of orientation of blank. It incrementally alters the orientation of blank by 1 degree and then calculates the material utilization of sheet at each angle till the blank has been rotated by 180 degree from its initial position. The orientation that has the maximum utilization ratio is the optimal. The outputs of this module are stored automatically in a data file labelled as MAXUTL.DAT.

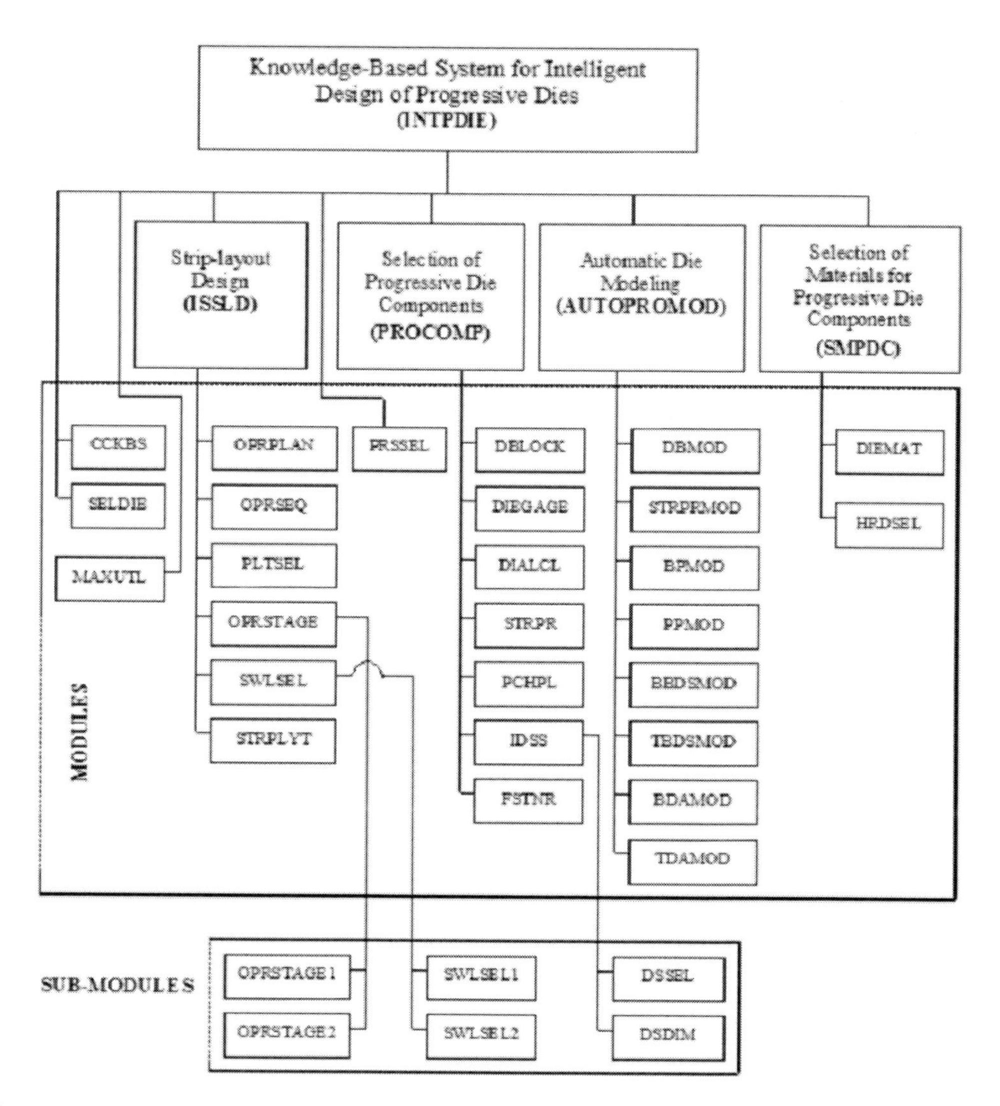

Figure 4. Organization of the system INTPDIE.

Sub-system ISSLD is developed for intelligent design of strip-layout. This sub-system comprises of six modules. The first module OPRPLAN determines the type of sheet metal operations required to manufacture the part. The next module OPRSEQ of the sub-system determines the sequencing of recommended sheet metal operations. The module PLTSEL is developed for selection of proper piloting scheme for positioning the strip accurately in each station of progressive die. The next module OPRSTAGE is developed for imparting expert advices for number of stations required and preferred staging of operations on progressive die. This module has two sub-modules namely OPRSTAGE1 and OPRSTAGE2. The first sub-module is designed to impart general expert advices for staging of operations. The second sub-module OPRSTAGE2 is developed for deciding number of stations required and operations to be accomplished at each station of progressive die. Module SWLSEL determines the proper size of sheet metal strip. This module has been structured into two sub-modules, namely SWLSEL1 and SWLSEL2. The first sub-module SWLSEL1 is developed for selection of strip width and the later for selection of feed distance. The module STRPLYT models the strip-layout automatically in the drawing editor of AutoCAD. The execution of the sub-system is shown through a flow chart in Figure5.

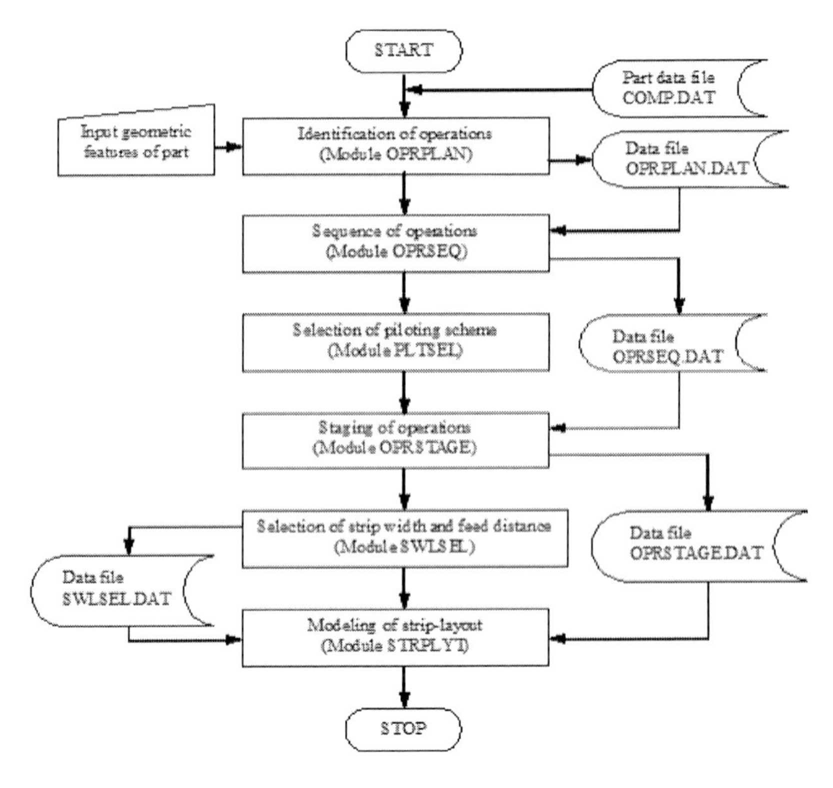

Figure 5. Execution of the sub-system ISSLD.

Module PRSSEL is constructed for assisting the user in the selection of suitable type of press machine for carrying out required sheet metal operations on progressive die. The module displays the minimum force required for carrying the required sheet metal operations and identifies the suitable alternative press machines having sufficient tonnage capacity. The module also calculates and displays the unit cost of part on each candidate press machine. Finally, it identifies the press machine on which the sheet metal part can be produced most economically.

The sub-system PROCOMP is designed for selection of progressive die components. Since the progressive die has several components, therefore the sub-system PROCOMP is structured into following modules –

1. Module DBLOCK for selection of size of die block.
2. Module DIEGAGE for selection of size of die gages and distance between die gages.
3. Module DIALCL for selection of proper die angle, die land and cutting clearance.
4. Module STRPR for selection of type and size of stripper and stripper plate.
5. Module PCHPL for selection of punch details, punch plate and back plate.
6. Module IDSS for selection of type and size of die-set. This module has two sub-modules, namely -
   (i) DSSEL for selection of type of die-set, and
   (ii) DSDIM for selection of size of die-set
7. Module FSTNR for selection of fasteners (bolts and dowels).

The suitable sizes of die components as recommended by modules are stored automatically in various output data files. Size of stripper plate, punch plate, back plate, die-set and fasteners also depend on the size of die block, therefore the output data file DBLOCK.DAT (Size of die block) is recalled during the execution of modules developed for the selection of these die components. All the output data files generated during the execution of modules of the sub-system PROCOMP are also utilized during automatic modeling of die components and die assembly. The execution of the sub-system PROCOMP is shown through a flow chart in Figure 6.

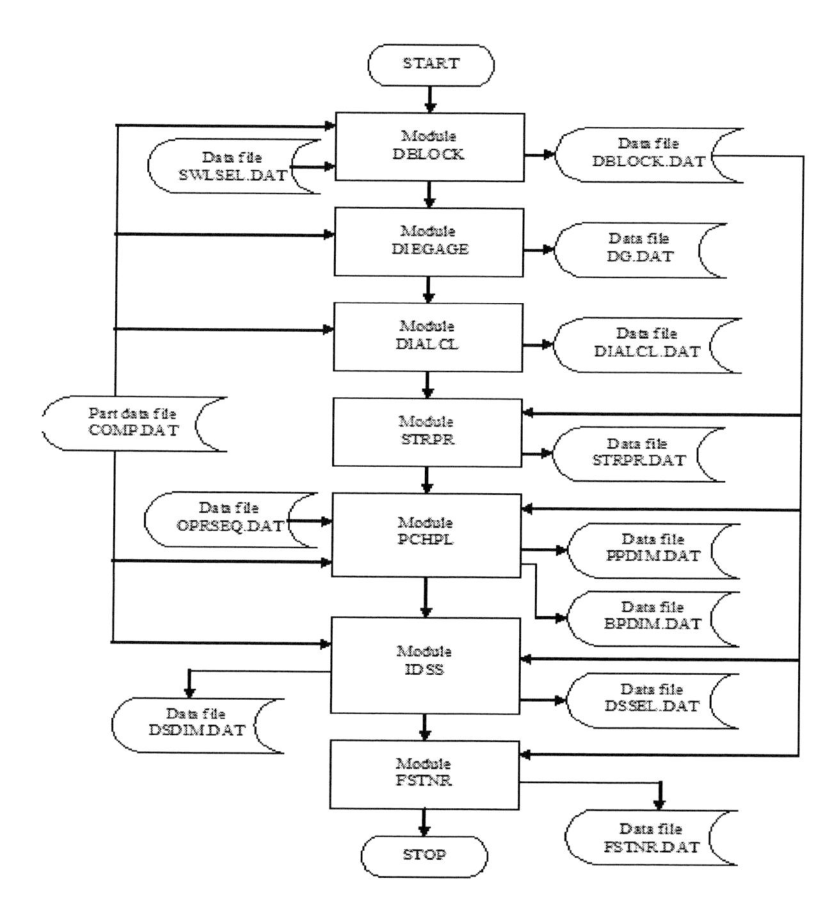

Figure 6. Execution of the sub-system PROCOMP.

The sub-system AUTOPROMOD is constructed for automatic modeling of progressive die components and die assembly in the drawing editor of AutoCAD. This sub-system works in integration with earlier modules developed for selection of progressive die components. The sub-system PROCOMP is structured into eight modules, namely DBMOD, STRPRMOD, BPMOD, PPMOD, BBDSMOD, TBDSMOD, BDAMOD and TDAMOD. The first module DBMOD of the sub-system is designed for automatic modeling of die block. The next module STRPRMOD is constructed for automatic modeling of stripper plate in the drawing editor of AutoCAD. The module BPMOD models the orthographic views of back plate in the drawing editor of AutoCAD. The module PPMOD models the punch plate. Modules BBDSMOD and TBDSMOD are constructed respectively for automatic modeling of bottom bolster and top bolster of die-set of progressive die. The module BDAMOD is developed for automatic modeling of lower or bottom portion of progressive die assembly. The last module TDAMOD models the top portion of progressive die assembly. The user has the option to modify these drawings through editing respective data files of die components or using AutoCAD commands.

The sub-system SMPDC is developed for selection of materials for progressive die components. It has two modules, labelled as DIEMAT and SELHRD. The first module is constructed for selection of materials for progressive die components and the second module SELHRD is designed for determination of hardness range of materials selected for punches and die/inserts of progressive die.

4.2. Validation of the Proposed System INTPDIE

The developed system INTPDIE is tested for different types of sheet metal parts for the problem of design of progressive die. A sample of typical prompts, user responses and the recommendations obtained by the user during the execution of the system for one example component (Figure 7) is given through Table 4. The strip-layout generated by the proposed system is shown in Figure 8 and the front and top view of bottom die assembly is shown in Figure 9.

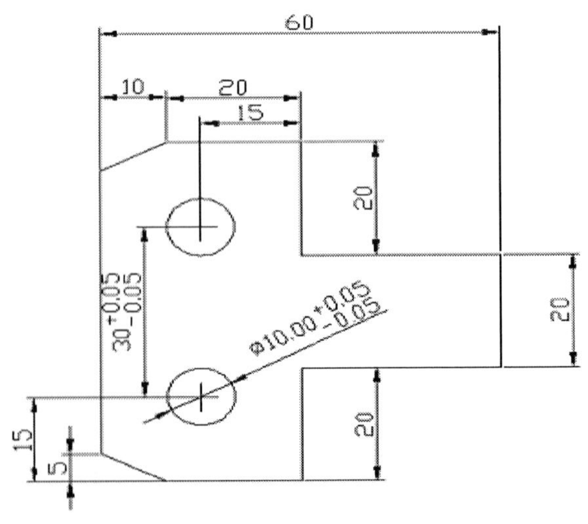

Figure7. Example Component (All dimensions are in mm).

The recommendations obtained by the system and drawings generated are found to be reasonably close to those actually used in industry (Tristar India Pvt. Ltd., New Delhi, India) for the example component. The system takes only 30 minutes to complete the design process of progressive die and generates the outputs in form of detailed drawings of strip-layout and die components; and assembled drawings of die.

Table 4. A sample of typical prompts, user responses and expert advices generated during execution of the system INTPDIE for example component (Figure7)

S. No.	Prompt	Example data entry	Advice to the user
1	Please enter sheet material	Brass	
2	Please enter sheet thickness in mm	1.0	
3	Please enter shape of holes on the part	Round	
4	Please enter diameter of round hole on part in mm.	10.0	Accept the diameter of hole
5	Please enter minimum hole pitch in mm.	30.0	Accept the hole pitch
6	Please enter the shape of component edge	Straight	
7	Please enter the maximum dimension of the component in mm (length/width)	60.0	Set lead end scrap web allowance in mm = 1.5 Set front/back scrap web allowance in mm = 1.5
8	Enter number of operations required	3	Design a Progressive die for manufacturing the example component
9	Are holes located on opposite sides of the part?	Yes	Select the two largest holes located on opposite sides of part for piloting
10	Is the center-to-center distance between holes has tolerance range within ± 0.05 mm?	Yes	Pierce these holes at the same station
11	Enter no. of notch in the part	2	
12	Enter no. of holes in the part	2	Number of stations required = 4 Preferred staging is given as under: First station: Piercing Second station: Notching and Piloting Third station: Notching and Piloting Fourth station: Parting off
13			Please enter command DSDIM for invoking the second Sub module DSDIM of the system IDSS
14	DSDIM		Place die parallel in the 4 pillar die-set with pillar diameter 40 mm and bolster dimensions in mm as: length = 450.0, width = 400.0, and height = 45.0

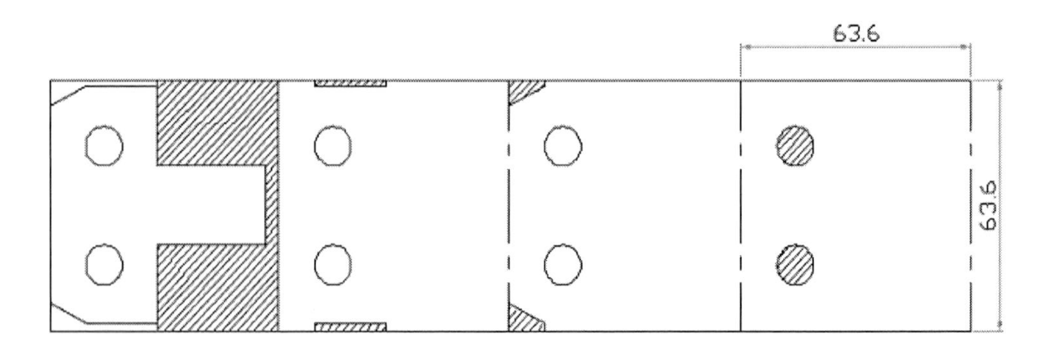

Figure 8 Strip-Layout Modeled by the System ISSLD for Example Component.

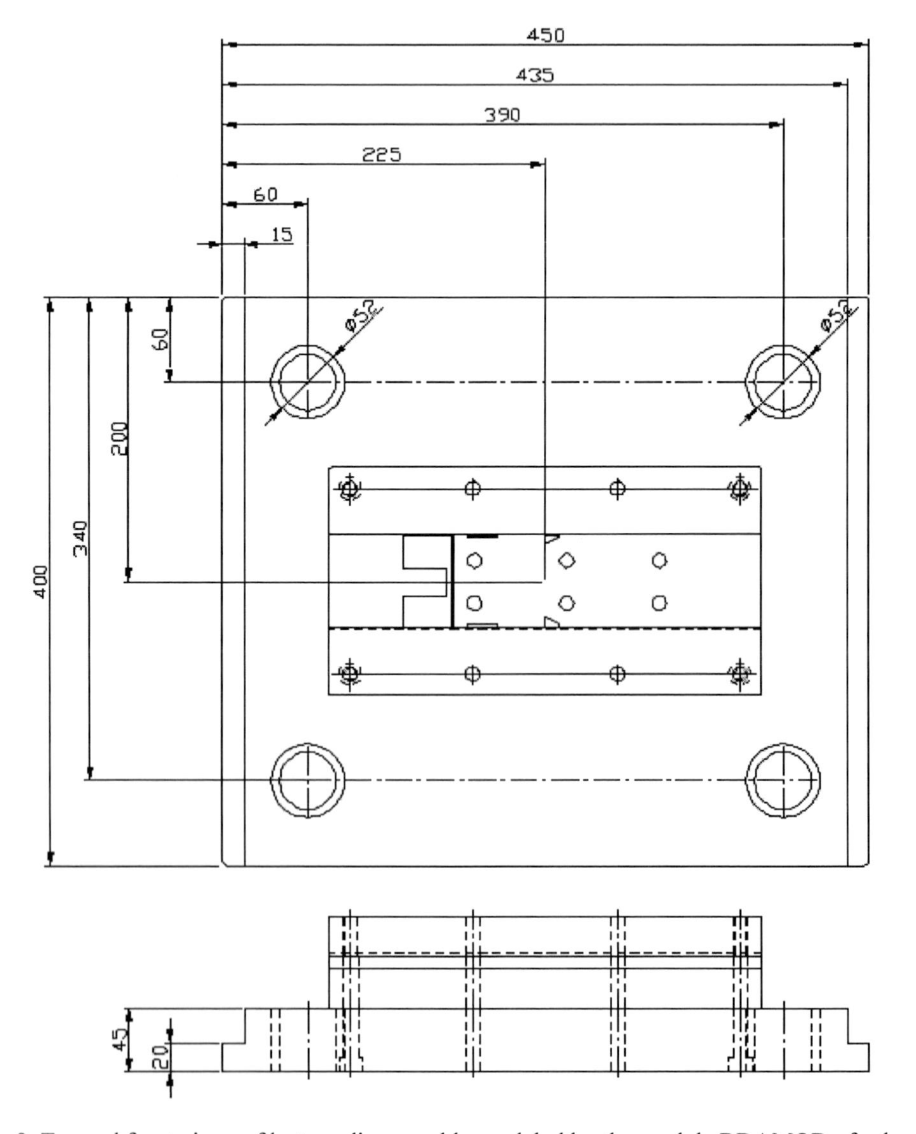

Figure 9. Top and front views of bottom die assembly modeled by the module BDAMOD of sub-system AUTOPROMOD.

4.3. Scope of Further Research Work

The present investigation contributes to the growing field of KBS for intelligent design of progressive die through the development of modules for checking of part design features, selection of type of die, strip-layout design, selection of die components, automatic modeling of die components and die assembly of progressive die; and selection of materials for progressive die components. Although the KBS developed in the present work is ready for use in the sheet metal industries, there is scope for further improvements. Few suggestions for further research are given below.

1. A system can be developed for analysis of die design data from strength point of view using Finite Element Analysis (FEA) or some other suitable approach.
2. Investigations can be made for prediction of die stresses and die life.
3. A post processor program can be developed, which is capable of connecting the outputs of the proposed KBS to the Computer-Aided Manufacturing (CAM).
4. Investigations can be made for the development of KBS for intelligent design of other types of dies also such as drawing die, bending die, forming die, combination die and compound die.

CONCLUSION

Traditionally, process planning of stamping operations and design of metal stamping dies require experienced die designers; involve numerous calculations, and hence time-consuming tasks. But with the advancement in the field of AI around 1980s, these are being carried out using various AI techniques. But most of the systems developed using AI techniques are having limitation in extraction and representation of part feature data in more interactive format for displaying output. Further, most of the systems are developed for specific application. Very few systems are developed for design of multi-operations dies and even these are not capable to fully automate the die design process. Therefore, there is need to develop an intelligent system by combining some suitable AI technique and CAD system for manufacturability assessments/reasoning, concurrent planning and quick design of multi-operation dies. The system must have rich knowledge-base comprising knowledge of experienced die designers and process planners, must be interactive and user friendly and have low cost of implementation.

In the present chapter, sheet metal operations, type of press tools and various AI techniques are described briefly. Research work done by worldwide researchers in the area of applications of AI techniques to metal stamping dies are critically reviewed and salient features of their work is presented in tabular form. Further, the research investigation done by the author for development of an intelligent system for automation of design of progressive dies is presented at length. The developed system is capable to provide assistance to the process planners and die designers of small and medium scale sheet metal industries. However, further research efforts are required to develop such intelligent systems for design of other types of multi-operation stamping dies.

REFERENCES

[1] Nee, A.Y. C. *J. Mech. Working Technol.* 1989, *19*, 11-21.

[2] Barr, A.; Feigenbaum, E. A. The hand book of artificial intelligence; William Kauf Manu, Loss Altos, California, 1981, Vol. 1.

[3] Pham, D. T. Artificial Intelligence in Engineering; 1st Conference on Applications of AI Techniques in Engineering, 5-7 Oct., 1994, Naples: 4-37.

[4] Makinouchi, A. *J. Mater. Process. Technol.* 1996, *60*, 19-26.

[5] Prasad, Y.K.D.V.; Somasundaram, S. *Comput. and Control Eng. J.* 1992, *3*, 185-191.

[6] Lazaro, A.D.S.; Engquist, D.T.; Edwards, D.B. *Concurrent Engineering: Res. and Applications,* 1993, *1*, 117–123.

[7] Shpitalni, M. ; Saddan, D. *Ann. CIRP*. 1994, *43*, 23–26.

[8] Meerkamm, H. *Proc. of Product Modeler,* Erlangen, University of Erlangen, 1995, pp. 25–47.

[9] Lee, R.S.; Chuang, L.C.; Yu, T.T.; Wu, M.T. "Development of an assessment system for sheet metal forming," *Proc. of the Int. Conf. on precision Engrs. (2nd ICMT),* Trends and innovations in precision manuf. Technol., Singapore, 1995, pp. 515-518.

[10] Mantripragada, R.; Kinfzel, G.; Altan, T. *J. Mater. Process. Technol.* 1996, *59*, 241–248.

[11] Yeh, S.; Kamran, M.; Terry, J. M. E.; Nnaji, B.O. *IIE Trans*. 1996, *28*, 1–10.

[12] Radhakrishnan, R.; Araya, A.; Kamran, M.; Nnaji, B.O. *J. Manuf. Sys*. 1996, *15*, 179-189.

[13] Gupta, S. K.; Bourne, D. A.; Kim, K.H.; Krishnan, S. S. *J. Manuf. Sys.* 1998*, 17*, 338–360.

[14] Xie, S.Q.; Tu, Y.L.; Liu, J.Q.; Zhou, Z. D. *Int. J. Prod. Res*. 2001, *39*, 1095–1112.

[15] Tang, D.B.; Zheng, L.; Zhizhong, L.; Chin, K. *Int. J. Manuf. Technol.* 2001*, 18*, 193-200.

[16] Ramana, K.V.; Rao, P.V.M. *Int. J. Prod. Res*. 2005, *43*, 3889-3913.

[17] Giannakakis, T.; Vosniakos, G. C. *Int. J. Adv. Manuf. Technol.* 2008, 658–670.

[18] Shukor, S.A.; Axinte, D. A. *Int. J. Prod. Res*. 2009, *47*, 1369- 1390.

[19] Lin, Z. C.; Hsu, C.Y.; Yao, K.H. *J. Chin. Soc. Mech. Engrs.* 1989, *10*, 101–120.

[20] Sitaraman, S. K.; Kinzel, G. L.; Altan, T. *J. Mater. Process. Technol.* 1991, *25*, 247-271.

[21] Lee, B. H.; Lim, B. S.; Nee, A.Y.C. Computer-Aided Prod. Engineering, 1991, 8(1), 19-33.

[22] Lee, I. B. H.; Lim, B. S.; Nee, A.Y.C. *Int. J. Prod. Res. 1993*, *31 (2)*, 251-278.

[23] Duffy, M. R.; Sun, Q. *J. Mater. Process. Technol.* 1991, *28*, 221-227.

[24] Li, Z.; Li, J.; Li, Z.; Wang, G.; Xiao, J. "CAD/CAM of progressive dies for household electronic appliances" *Proc. of Int. Conf on Adv. Technol. and Machinery in Metal Forming*, Wuhan China, 1992, pp.71–77.

[25] Nee, A. Y. C.; Foong, K. Y. *J. Mater. Process. Technol*. 1992, *29*, 147–158.

[26] Ismail, H. S.; Huang, K. *Proc. Institution of Mech. Engrs. Part B: J. Engg. Manufact.* 1993, 207, 117- 127.

[27] Ismail, H. S.; Hon, K. K. B.; Huang, K. *Ann. CIRP*. 1995, *44 (1),* 91-96.

[28] Ismail, H. S.; Chen, S. T.; Hon, K. K. B. *Int. J. Machine Tools and Manufact.* 1996, *36*, 367-378.

[29] Huang, K.; Ismail, H. S.; Hon, K. K. B. *Proc. Institution of Mech. Engrs. Part B: J. Engg. Manufact.* 1996, *210*, 367–376.

[30] Tisza M. *J. Mater. Process. Technol.* 1995, *53,* 423–432.

[31] Cheok, B. T.; Foong, K. Y. *Proc. Institution of Mech. Engrs. Part B : J. Engg. Manufact.* 1996, *210*, 25-35.

[32] Cheok, B. T.; Nee, A. Y. C. *Artificial Intelligence for Engineering Design, Analysis and Manufacturing,* 1998, *12*, 405-418.

[33] Esche, S. K.; Khamitkar, S; Kinzel, G.; Altan, T. *J. Mater. Process. Technol.* 1996, *59*, 24–33.

[34] Ong, S.K. ; De Vin, L.J. ; Nee, A.Y.C. ; Kals, H.J.J. *J. Mater. Process. Technol.* 1997, *69*, 29–36.

[35] Park, S. B. *J. Mater. Process. Technol.* 1999, *88*, 216–221.

[36] Choudhary, S.; Allada, V. *Int. J. Adv. Manufact. Technol.* 1999, *15*, 356-365.

[37] Singh, R.; Sekhon, G. S. *J. Mater. Process. Technol.* 1999, *86*, 131-138.

[38] Pilani, R.; Narasimhan, K.; Maiti, S. K.; Singh, U. P.; Date, P. P. *Int. J. Adv. Manufact. Technol.* 2000, *16*, 370-375.

[39] Caiyuan, L.; Jianjun, L.; Jianyong, W.; Xiangzhi, X. *Int. J. Prod. Res.* 2001, *39/18*, 4133–4155.

[40] Choi, J. C.; Kim, C. *J. Mater. Process. Technol.* 2001, *110*, 36-46.

[41] Shi, X.; Chen, J.; Peng, Y. ; Ruan, X. *Int. J. Adv. Manufact. Technol.* 2002, *19*, 898–904.

[42] Tor, S. B.; Britton, G. A.; Zhang, W. Y. *J. Computing and Information Sci. in Eng., ASME* 2003, *3*, 355-362.

[43] Vosniakos, G. C.; Segredou, I.; Giannakakis, T. *J. Intell. Manufact.* 2005, *16*, 479–497.

[44] Dequan, Y.; Rui, Z.; Jun, C.; Zhen, Z. *Int. J. Manufact. Technol.* 2006, *29*, 663-669.

[45] Chan, W.L.; Fu, M.W.; Lu, J. *Eng. Appli. Artificial Intell.* 2008, 21, 1170– 1181.

[46] Chi Z, "Mining rules from stamping die designs," *3rd Int. Conf. Innovative Compu. Information and Control,* Dalian, 2008, pp.74 – 74.

[47] Zhibing, Z.; Yuqi, L. ; Ting, D.; Zhigang, L. *J. Mater. Process. Technol.* 2008, *205*, 425–431.

[48] Chu, C.Y.; Tor, S. B.; Britton, G. A. *Int. J. Prod. Res.* 2008, *46*, 2965-2988.

[49] Ghatrehnaby, M.; Arezoo, B. *J. Mater. Process. Technol.* 2009, *209*, 525–535.

[50] Yi-Lung Tsai, Yi-Lung; You, Chun-Fong; Lin, Jhen-Yang; Liu, Kun-Yu. *Computer Aided Design and Applications*, 2010, *7(1)*, 75-87.

[51] Kumar, S. *A contribution to the development of knowledge-based system for intelligent design of progressive dies.* PhD thesis, 2006, M. D. University, Rohtak, Haryana, India.

In: Artificial Intelligence
Editor: Brent M. Gordon, pp. 97-114
ISBN 978-1-61324-019-9

Chapter 4

STRUCTURAL FEATURES SIMULATION ON MECHANOCHEMICAL SYNTHESIS OF AL_2O_3-TIB_2 NANOCOMPOSITE USING ANN WITH BAYESIAN REGULARIZATION AND ANFIS

A. Ghafari Nazari*[*1], V. R. Adineh[2] and A. Alidoosti[2]
[1]Materials Engineering Department, Islamic Azad University, Southern Branch, Tehran, Iran
[2]Department of Mechanical Engineering, Saveh Branch, Islamic Azad University, Saveh, Iran

ABSTRACT

In this study, structural features ofAlumina-Titanium diboridenanocomposite (Al_2O_3-TiB_2) were simulated from the mixture of titanium dioxide, boric acid and pure aluminum as raw materials via mechanochemical process using Artificial Intelligence approaches (AI). The phase transformation and structural evolutions during the mechanochemical process were characterized using X-Ray powder Diffractometry(XRD). For better understanding the refining crystallite size and amorphization phenomenaduring the milling, XRD data were modeled using Adaptive Neuro Fuzzy Inference System (ANFIS) and Artificial Neural Networks (ANN). Results show that ANN has better performance compared to ANFIS. The best predictor is then selected for simulation of crystallite size, interplaner distance, amorphization degree, and lattice strain. Furthermore, the simulated results are compared with experimental results.A good agreementbetween the experimental results and simulation ones were achieved.

Keywords: Crystalline state; Mechanical alloying; X-ray diffraction.

* E-mail: alighafarinazari@yahoo.de

1. INTRODUCTION

Titanium diboride (TiB_2) is an attractive combination of high Vickers hardness, electrical conductivity, excellent chemical resistance to molten nonferrous metals and relatively low specific gravity [1, 2]. TiB_2has some mechanically poor, such as fracture toughness and impact strength, these properties are improved by making composite. Therefore, TiB_2-Al_2O_3 composite is useful for a variety of applications including cutting tools and wear-resistant substrates [3-5]. Recently, mechanical activation and mechanical milling have been extensively implemented for preparing and synthesizing nanocomposite powders [6].

Mechanochemical is concerned with the physical and chemical change of materials caused by mechanical energy. During the milling repeated welding and fracturing of powder particles increases the area of contact between the reactant powder particles due to a reduction in particle size and allow fresh surfaces to repeatedly come into contact; this causes the reaction to proceed without the necessary diffusion through the product layer which enhances the formation of new compounds, amorphization of the crystalline structures, phase transformation and formation of chemical reaction [7-9]. There is a complicated and non-linear relationship among the operational parameters such as time and speed of milling, Ball-to-Powder weight Ratio (BPR), raw materials, number and diameter of balls, and atmosphere of milling on the properties of product [7].X-Ray Diffractometry(XRD) is a powerful technique for characterizing solids which is widely applied in identifying the crystalline solid phases and offers a unique advantage in the quantitative analysis of mixtures [10].

In this research, the experimental studies have been conducted to examine the effects of significant parameters on the mechanochemical synthesis in TiB_2-Al_2O_3nanocomposite. This study evaluates the feasibility of using Artificial Intelligence (AI) methods in recognizing peak-shaped signals in the analytical data. It was concluded that AI is superior to the conventional classifiers (e.g. multiple linear regressions)in classifying patterns in which the input is noisy and the system is not well defined[11]. Besides, AI approaches are non-linear estimators which can establish more sophisticated responses. This research demonstrates the application of AI methods consists of Artificial Neural Networks (ANN) and Adaptive Neuro Fuzzy Inference System (ANFIS) as well, for simulating the aforesaid structural features. Several studies were conducted for usage of AI methods in this field [11-14]. All of them are concerned with only three pairs of XRD patterns for predicting chemical compositions, while this research utilizes more than 1500 ones in order to simulate the structural characteristics.

There are various challenges in the AI modeling such as setting the parameters of ANFIS modeling, selecting the appropriate input variables for ANFIS training, choosing the optimum topology for ANN, and so forth. To cope with these problems, as it will be discussed later, several strategies are utilized. For instance, to do input selection, method of Jang [15] is utilized. Moreover, for optimizing the ANN, Taguchi method was used in which non-numerical parameters, such as types of functions, are optimized versus other statistical methods [16].

This chapter is organized as follows. Section 2 describes the materials and methods used in this study. Modeling of XRD intensity is explained in section 3. Section 4 discusses about obtained results and demonstrates the simulation of crystalline properties and section 5 concludes the chapter.

2. EXPERIMENTALPROCEDURES

Titanium dioxide (TiO_2, Merck), aluminum powder(Merck) and boric acid (H_3BO_3, Merck) were utilized as raw materials. The mixture of raw materials was prepared according to the stoicheiometry given be following reactions:

$$2H_3BO_3 \rightarrow B_2O_3 + 3H_2O \tag{1}$$

$$3TiO_2 + 3B_2O_3 + 10Al \rightarrow 3TiB_2 + 5Al_2O_3 \tag{2}$$

A planetary ball mill was applied in milling experiments. Preliminary experiments indicated that the following conditions were appropriate: 600 rpm and 5 milling balls with a diameter of 20mm and made of high Cr-steel in a high Cr-Steel milling chamber giving a BPR of 20:1. The milling runs were conducted in an argon atmosphere.

The XRD scans were performed on a Philips wide angle X-ray powder diffractometry with an X-ray generator (1840 PW). A copper target X-ray tube was operated at a power of 40 kV, 20 mA to get the idea of the relative crystalline of the composite at 2 mm slits at a step size of 0.05° and a scan rate of 3 sec. The crystalline and amorphous portions were determined by arbitrary units. The degree of amorphization (Φ) was measured using the following relationship:

$$\phi = \frac{I_a}{I_a + I_c} \tag{3}$$

whereI_a and I_c are the integrated intensity of the amorphous and crystalline region respectively [13], where, the crystallite size (p), the interplaner distance (d) and the lattice strain (η) were calculated as follows:

$$p = \frac{k\lambda}{\beta \cos\theta} \tag{4}$$

$$d = \frac{\lambda}{2\sin\theta} \tag{5}$$

$$\eta = \frac{\sqrt{\beta}}{|4\tan\theta|} \tag{6}$$

where β (in radian), K, λ, d,and θ are the half-height width of the crystalline peak which is the difference in integral profile width between the standard and unknown sample, the shape coefficient, the wavelength of the X-ray used, crystallite size, and the Bragg angle, respectively.

3. Modeling Intensityin XRD

3.1. Pre-Processing of the Data

Before the AI can be trained and the mapping could be learnt, it is important to process the experimental data into patterns. Thus, training pattern vectors are formed; each pattern is formed with an input condition vector and the corresponding target vector [17]. For increasing the accuracy and processing velocity of network, the data were cleaned. In other words, XRD of unmilled materials and intensity at 10-24° 2θ for the milled samples were omitted since they are not significantly different. The input pattern vectors are then formed, comprising 8400 pairs of input/output ones for training of ANFIS and ANN. That is, the total number of training patterns were 11200 ones which after cleaning process they were reduced to 8400 ones.

Also, the scale of the input and output data is an important matter to consider, especially when the operating ranges of the process parameters are different. The scaling or normalizing ensures that the AI will be trained effectively, with no particular variable significantly skewing the results [17]. The scaling is performed by mapping each variable at a boundary of [0, 1]. The milling time was normalized by equation 7 and XRD angle and XRD intensity were normalized according to equation 8 [18].

$$t_n = \frac{\log(t)}{\log(40)} \tag{7}$$

$$X_n = \frac{X - X_{\min}}{X_{\max} - X_{\min}} \tag{8}$$

whereX_{min} and X_{max} are minimum and maximum of all data sets and X_n is normalized parameter. In equation 7, 40 is the maximum value of milling time.

3.2. ANFIS

An adaptive network, as its name implies, is a network structure consisting of nodes and directional links through which the nodes are connected. Moreover, parts or all of the nodes are adaptive, which means each output of these nodes depends on the parameters pertaining to this node and the learning rule specifies how these parameters should be changed to minimize a prescribe error measure [19].

There are two adaptive layers in ANFIS architecture, namely the first layer and the fourth layer. In the first layer, there are three modifiable parameters, which are related to the input Membership Functions (MF). In the fourth layer, there are also three modifiable parameters pertaining to the first order polynomial [20].

Although Back Propagation or steepest descent method can be applied for the parameter identification of adaptive network, but the convergence speed of this simple method is very slow and easily run into local minimum [21]. Since the output of an adaptive network is linear combination of some network parameters, linear least-squares method can be used to identify linear parameters, that the least square method can be combined with the steepest descent method to identify the system's parameters.Since it has been proven that hybrid algorithm is highly efficient in training the ANFIS [19], hybrid method was selected as the learning algorithm.

Considering and applying two inputs in all the data sets, the following are designed: the output intensity in XRD derived from different the milling time, networks with two inputs, milling time and XRD angle, and one neuron in output layer. Figure 1 shows the considered ANFIS topology, input, and output parameters.

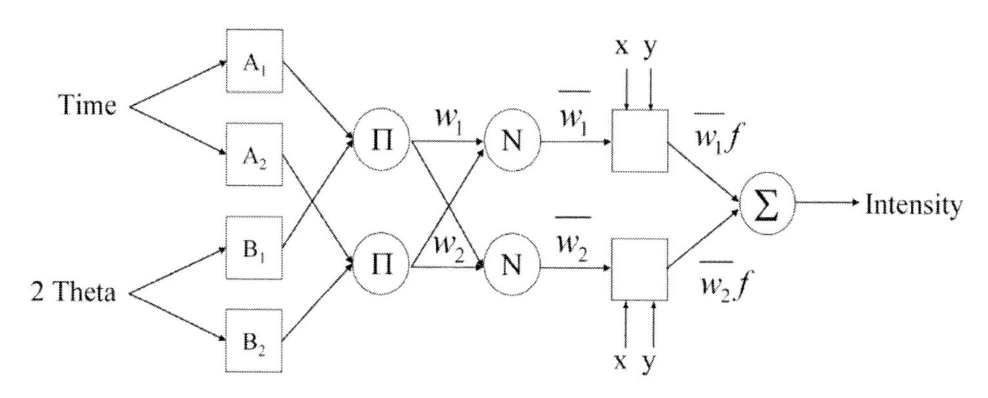

Figure 1.The ANFIS topology.

3.3. ANN with Bayesian Regularization with Full Sampling

ANN is an information-processing program inspired by mammalian brain processes. It is composed of a number of interconnected processing elements analogous to neurons. Each neuron has a transfer function. The neurons are linked via connections analogous to synapses, each link having its own transfer function or "weight". In biological systems, learning involves adjustments to the weights of the synapses. This is true of ANN as well in which a training algorithm adjusts connection weights iteratively so as to maximize ANN response accuracy during the training. During training process, data sets of specific input variables and the specific answer that should be arrived at from this input [17]. Thus the training algorithm gives a set of input data to the ANN and checks the ANN output for the desired results. If the actual/desired output match is below a specified level of accuracy, the training program will modify the connection weight; then it will test the same input again. If the actual/desired output match improves, the same type of modification (simply presenting this idea) will be followed. If the match deteriorates, the type of modification will be reversed. This process continues until the desired accuracy is attained.

The Feed Forward Back Propagation (FFBP) [22]is a multiple-layer network with an input layer, an output layer and some hidden layers between the input and output layers. The input–output relationship between each node of a hidden layer can be written as:

$$a_k = f\left(\sum_{i=1}^{k} w_{ki} p_i + b_k\right) \tag{9}$$

wherep_i is the output from the$i$th node of the previous layer, $w_{ki}$is the weight of the connection between the ith node and the currentnode, and b_k is the bias of the current node. f is an activation function which can be considered log-sigmoid, tan-sigmoid or liner as follows:

$$f(z) = \frac{1}{1+e^{-z}} \text{(LOGSIG)} \tag{10}$$

$$f(z) = \frac{2}{(1+e^{-2z})-1} \text{(TANSIG)} \tag{11}$$

$$f(z) = z \text{ (PURELIN)} \tag{12}$$

Before practical implementation, the network has to be trained so that the free parameters or connection weights are determined and the mapping between inputs and outputs is accomplished. The training method is called back-propagation, a supervised learning technique, which generally involves two phases through different layers of the network: a forward phase and a backward phase. In the forward phase, input vectors are presented and propagated forward to compute the output for each neuron. During this phase, synaptic weights are calculated and the Mean Square Errors (MSE)of all the patterns of the training set is calculated by;

$$MSE = \frac{1}{n}\sum_{i=1}^{n}(t_k - a_k)^2 \tag{13}$$

where n is the total number of training patterns, t_k is the target value, and a_k is the network output value. The backward phase is an iterative which is error reduction performed in a backward direction from the output layer to the input layer. The gradient descent method, adding a momentum term, is usually used to minimize the error (MSE) as fast as possible [17, 22]. These two phases are iterated until the weight factors stabilize their values and the MSE reaches a minimum or an acceptably small value.

FFBP were used with Bayesian Regularization (BR) training algorithm for training the input patterns.In most articles [10, 11, 13, 14, 17, 23-25], data sets have been divided to train and test sets because of overfitting problem; that is ANN could not be generalized. This solution for overfitting was called early stopping [22].Early stopping is the only solution for overfitting at ANFIS. But at ANN another solution for this situation is BR. BR is desirable for determining the optimal regularization parameters in an automated fashion [26, 27]. In comparison, BR is easier and approximately 5 times more accurate than early stopping [22].

3.4. Optimizing ANN by Taguchi Method

FFBP has a number of parameters such as the number of hidden layers and their neurons, typeof transfer function in hidden layers and output layer, and type of learning functions. There are several formulas based upon rule of thumb for these selections which mostly depend on the desired approximation order, number of inputs and amount of accuracy and importance required for the estimation. Nevertheless, there is no explicit formulas available for choosing the number of necessary hidden units which do not need any quantities of specific functions for acceptable prediction [28]. As a result, selecting the best ANNby trial and error is proposed by several textbooks [22, 27] and papers[13, 24]. In this paper, the taguchi method is utilized for better selection of parameters.

Taguchi method is a methodology which chooses the most suitable combination of the levels of controllable factors by applying Signal to Noise ratio (S/N) table and orthogonal arrays against the factors that form the variation and are uncontrollable in the process. Hence, it tries to reduce the variation in product and process to the least. Taguchi uses statistical performance measure which is known as S/N that takes both medium and variation into consideration [29]. Taguchi method is a robust design for optimizing numericaland non-numerical parameters with no limitation [16].

The selected factors are shown in Table 1 along with their levels. These parameters were chosen after some trials. Based on this selection, Taguchi method proposes L_8 array (8 trials) [16]. Additionally, each of the trials was run twice. This optimization's objective is to get the ANN with a minimum MSE in modeling the XRD patterns according to equation 13. S/N, the small the better, was calculated by equation 14.

$$S / N = -10 \, Log_{10} \frac{1}{n} \sum_{i=1}^{n} y_i^2 \tag{14}$$

Table 1.ANN factors with the levels selected for the Taguchi method

Factor	Level 1	Level 2	Level 3	Level 4
Topology (Neuron numbers)	2-10-5-1	2-20-10-1	2-30-15-1	2-40-20-1
First hidden layer transfer function	LOGSIG	TANSIG		
Second hidden layer transfer function	LOGSIG	TANSIG		
Learning function	LEARNINGD	LEARNINGDM		

wherey_i is MSE on every run and n is the number of rehearsal, i.e. n=2. The analysis of mean statistical approach is adopted herein to construct the optimal conditions. Additionally, the mean of the S/N ratio of each controllable factor at a certain level must be calculated. The mean of the S/N ratio of factor I in level i, $(M)_{\text{Factor=I}}^{\text{Level=i}}$ is given by:

$$(M)_{\text{Factor=I}}^{\text{Level=i}} = \frac{1}{n} \sum_{j=1}^{n_{Ii}} \left[(S / N)_{\text{Factor=I}}^{\text{Level=i}} \right]_j \tag{15}$$

In equation 15, n_{Ii} represents the number of appearances of factor I in the level i, and $\left[(S / N)_{\text{Factor=I}}^{\text{Level=i}} \right]_j$ is the S/N ratio of factor I in level i. By the same measure, the mean of the S/N ratios of the other factors in a certain level can be determined [30]. Thereby, the S/N response table is obtained, and the optimal conditions are established. Finally, the confirmation ANNs under these optimal conditions are carried out.

4. Results and Discussion

4.1. Experimental Data

In this section the experimental results are presented. Figure 2 shows the XRD patterns of the unmilled and milled samples for 1, 1.5, 2, 8, 20 and 40 h. Only sharp characteristic peaks of TiO_2, aluminum and H_3BO_3could be detected in the XRD pattern of unmilled and 1 h milled.

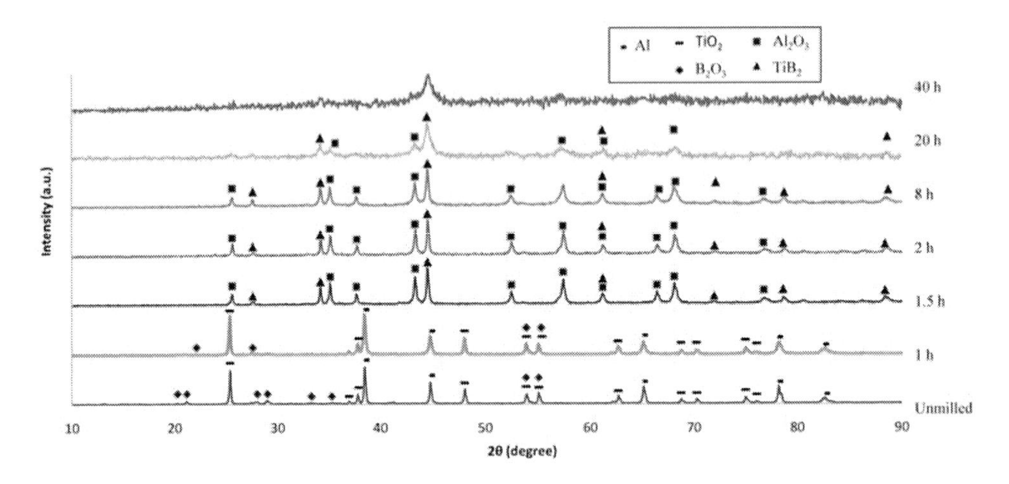

Figure 2.The XRD patterns.

According to the XRD profile (figure 2), the products after 1.5 h of milling were TiB_2 and Al_2O_3 phases. The peaks of TiB_2 and Al_2O_3 phases are approximately narrow and sharp especially in low angle ($2\theta < 60°$) diffraction. This suggests that the temperature during the reaction was sufficient to induce the crystallization of products. In addition, the absence of any reactant peaks confirms that the reaction is completed within 1.5 h.Sharifi et al [31] reports that the Al_2O_3-TiB_2 composite is produced after 60 h of milling.

To investigate the effect of milling time on the products, the milling runs were undertaken in the different milling times and simulated XRD patterns. It is clear that increasing the milling time beyond 2 h results in the broader peaks and less intense, as would be expected from the smaller crystallite size. Up to 2 h of milling, the major peaks for Al_2O_3 and TiB_2 decrease in intensity so that the peaks corresponding to Al_2O_3 completely disappear and the broadened peak for TiB_2phase appears after 40 h milling. The weakening of XRD peaks with increasing the milling time beyond 20 h is related to the amorphization of the crystalline phases.

4.2. Modeling by ANFIS

After cleaning and normalizing the gathered experimental data, there is a further processing for dividing the data into the train, validation and test sets. This is because ANFIS is very sensitive to the input variable selection for learning; that is, one of the great challenges in the modeling of nonlinear systems using ANFIS is selecting the important input variables from all possible input variables. Consequently, it is necessary to do input selection that finds the priority of each candidate inputs and uses them accordingly [15, 32]. for this purpose the Jang's method was selected [15]. The utilized input selection method is based on the assumption that "the ANFIS model with the smallest Root Mean Squared Error (RMSE) after one epoch of training has a greater potential of achieving a lower RMSE when given more epochs of training" [15]. Using a written code in the MATLAB, a set of train/validation/test sets was made randomly, and then they were trained for one epoch. The smallest one was used for further processing. These optimum sets are labeled as the best sets in tables 2 and 3, and the corresponding number is related to the values of RMSE after one epoch of training, for the best set amongst 1000 runs. In order to find the optimum ANFIS parameters for training, two different strategies were utilized:

- Various kinds of MFs with equal number of MFs (table 2)
- Various kinds of MFs with different number of MFs (table 3)

Besides, bell shape (gbell) and Gaussian (gaussmf) were selected as MFs types.

Table 2. Some ANFISs for the prediction of intensity in the XRD using first strategies

No.	No. MFs	Best Set	MFs Type	R^2			RSME		
				Train	Val.	Test	Train	Val.	Test
1	3-3	0.0803	gbelmf	0.514	0.400	0.451	0.075	0.093	0.087
			gaussmf	0.502	0.382	0.426	0.074	0.095	0.089
2	4-4	0.0761	gbelmf	0.542	0.457	0.420	0.070	0.087	0.093
			gaussmf	0.505	0.413	0.359	0.073	0.091	0.098
3	5-5	0.0777	gbelmf	0.612	0.533	0.510	0.066	0.084	0.082
			gaussmf	0.531	0.419	0.423	0.073	0.093	0.088
4	6-6	0.0754	gbelmf	0.581	0.480	0.466	0.068	0.088	0.086
			gaussmf	0.553	0.444	0.426	0.070	0.091	0.089
5	8-8	0.0754	gbelmf	0.629	0.562	0.551	0.064	0.078	0.082
			gaussmf	0.617	0.550	0.527	0.065	0.079	0.085
6	10-10	0.0742	gbelmf	0.698	0.609	0.531	0.059	0.073	0.083
			gaussmf	0.667	0.570	0.489	0.062	0.076	0.087
7	12-12	0.0740	gbelmf	0.705	0.573	0.565	0.058	0.078	0.078
			gaussmf	0.675	0.545	0.540	0.061	0.080	0.080

Table 3. Some ANFISs for the prediction of intensity in the XRD using second strategies

No.	No. MFs	Best Set	MFs Type	R^2			RSME		
				Train	Val.	Test	Train	Val.	Test
1	4-16	0.0751	gbelmf	0.690	0.641	0.590	0.059	0.069	0.078
			gaussmf	0.677	0.620	0.571	0.061	0.072	0.080
2	4-32	0.0702	gbelmf	0.769	0.717	0.671	0.051	0.065	0.071
			gaussmf	0.751	0.688	0.661	0.053	0.068	0.072
3	6-12	0.0757	gbelmf	0.685	0.599	0.545	0.060	0.074	0.082
			gaussmf	0.650	0.577	0.528	0.063	0.076	0.083
4	6-20	0.0673	gbelmf	0.749	0.587	0.654	0.053	0.076	0.071
			gaussmf	0.722	0.562	0.625	0.054	0.078	0.077
5	6-40	0.0593	gbelmf	0.845	0.696	0.732	0.043	0.067	0.060
			gaussmf	0.810	0.660	0.716	0.046	0.071	0.062
6	8-20	0.0682	gbelmf	0.757	0.623	0.628	0.053	0.074	0.071
			gaussmf	0.747	0.616	0.623	0.054	0.075	0.071
7	10-20	0.0676	gbelmf	0.716	0.617	0.637	0.058	0.076	0.070
			gaussmf	0.760	0.613	0.624	0.056	0.077	0.072

Although the results of table 3 are better than table 2, but if it does not follow that they are suitable for modeling and simulation of XRD patterns due to weak ability in modeling (low amounts of R^2 and RMSE).

4.3. Modeling by ANN

After cleaning and normalizing data similar to the ANFIS, various ANNs were designed and trained which results of some runs are presented in table 4. All of parameters of networks 1, 2 and 3 are identical except for the transfer function of output layer. It can be seen that TANSIG produces suite prediction compared to other ones. Nevertheless, by increasing the number of hidden layers and their neurons, the ability of network for prediction improves. But after network 9, by increasing the neurons of hidden layers to 60 and 30, the performance of ANN decreases.

Table 4. Some ANNs for the prediction of intensity in XRD

Network	Topology	Learning function	Transfer functions	Time	Epoch	SSE
1	2-20-1	LEARNGDM	TANSIG-PURELIN	2:59	749	39.4
2	2-20-1	LEARNGDM	TANSIG-LOGSIG	0:04	17	133
3	2-20-1	LEARNGDM	TANSIG-TANSIG	4:00	992	38.1
4	2-20-1-1	LEARNGDM	TANSIG-TANSIG	7:52	1839	32.3
5	2-20-3-1	LEARNGDM	TANSIG-TANSIG	36:50	6987	24
6	2-20-10-1	LEARNGDM	TANSIG-TANSIG	21:30	2574	17.7
7	2-20-10-1	LEARNGD	TANSIG-TANSIG	14:20	1778	15.1
8	2-20-10-1	LEARNGD	TANSIG-LOGSIG	24:02	2935	12.5
9	2-40-20-1	LEARNGD	TANSIG-LOGSIG	3:00:0	4831	6.63
10	2-60-30-1	LEARNGD	TANSIG-LOGSIG	1:10:0	393	18.0

In order to find the best simulator for XRD patterns, the best network of ANN (network 6 in the table 4) is compared with the best of ANFIS, i.e. bell shape of network 6 in the table 3. Results for 40 h milling are shown in figure 3.

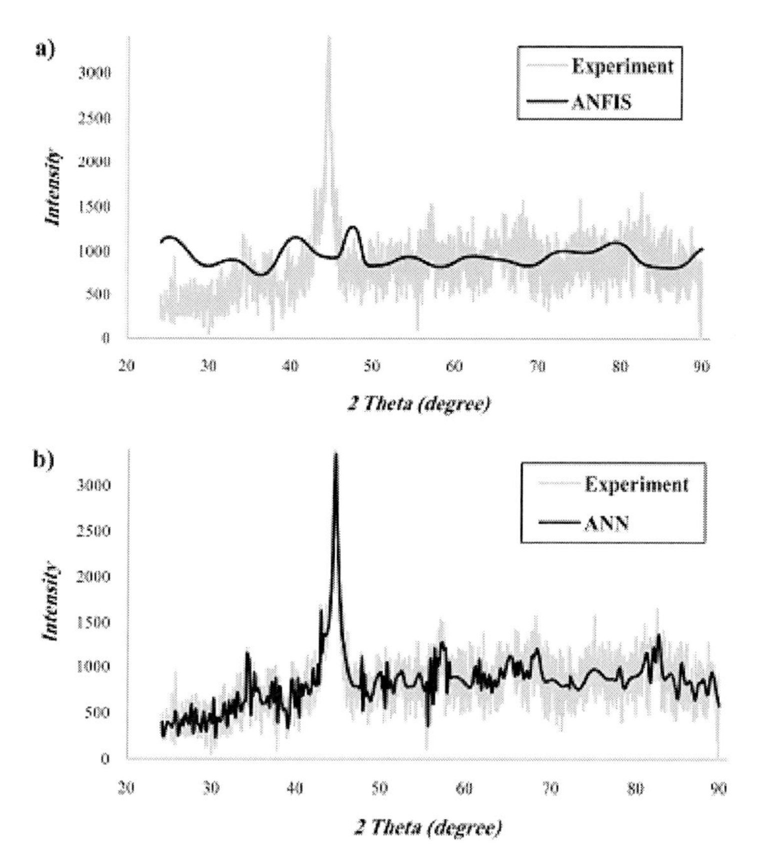

Figure 3. Graphically comparison of ANFIS (a) with ANN (b) based on the experiment results.

From the figure 3 it is established that ANN prediction is more accurate than ANFIS one. This result is contrary to those reported by Bose et al.[13] in which ANFIS has better performance for modeling the XRD patterns. It is notable that in the Bose et al. study, only the main peaks were selected instead of full peaks.

After presenting the appropriate factors and levels to the Taguchi method (according to the table 1), several networks were designed by L_8 array. In order to increase the accuracy of Taguchi method, networks were run twice. Networks with their results are presented in table 5. Time and epoch number of training is reported only for the first run. All of data sets were utilized for the prediction and calculation of SSE and R^2. Due to the nature of ANN [33], each run has a different MSE for every network. Substituting the number of ANN repetitions and results (i.e. the MSEs) into the equation 14, the S/N ratio of each network was determined (table 5). Subsequently, the values of the S/N ratio were substituted into the equation 15 and the mean of the S/N ratios of a certain factor in the *i*th level, $(M)_{\text{Factor=I}}^{\text{Level=i}}$ was obtained (table 6). In the table 6, the boldface numbers refer to the maximum value of the mean of the S/N ratios of a certain factor among their levels.

Table 5. Networks designed by Taguchi method and their results from twice running

Properties of networks by Taguchi design				Results of first run						
Net.	Topology	Learning function	Transfer function	Time	epoch	SSE	R^2	MSE 1	MSE 2	S/N
1	2-10-5-1	LEARNGD	LOGSIG-LOGSIG	3:31	760	35.6	0.820	0.0651	0.0638	23.8151
2	2-10-5-1	LEARNGDM	TANSIG-TANSIG	7:30	856	35.1	0.822	0.0647	0.0583	24.2108
3	2-20-10-1	LEARNGDM	LOGSIG-LOGSIG	57:40	7148	13.4	0.936	0.0400	0.0411	27.8394
4	2-20-10-1	LEARNGD	TANSIG-TANSIG	1:20:18	9600	13.6	0.935	0.0403	0.0439	27.5064
5	2-30-15-1	LEARNGD	LOGSIG-TANSIG	1:09:06	3955	10.0	0.950	0.0353	0.0351	29.0691
6	2-30-15-1	LEARNGDM	TANSIG-LOGSIG	1:10:30	3629	8.64	0.959	0.0321	0.0345	29.5455
7	2-40-20-1	LEARNGDM	LOGSIG-TANSIG	7:10:00	11543	7.23	0.969	0.0293	0.0289	30.7219
8	2-40-20-1	LEARNGD	TANSIG-LOGSIG	5:18:37	8545	6.45	0.969	0.0277	0.0277	31.1845
9	2-40-20-1	LEARNGDM	TANSIG-LOGSIG	5:11:05	8300	6.34	0.970	0.0274	0.0277	31.1847

Table 6. Mean effect of each level by Taguchi method calculations

Level	A	B	C	D
1	24.0129	27.8552	28.0961	27.8876
2	27.6729	28.1118	27.8709	28.0794
3	29.2950			
4	30.9532			

With regard to the table 6, the optimum topology, transfer functions of layers and learning function are 2-40-20-1, TANSIG, LOGSIG and LEARNGDM respectively. In the last column of the table 5, (i.e. network 9) results of optimized ANN are demonstrated. From the table 5 it is seen that the optimum network is very similar to the network 8. The only difference between them is the learning function. Moreover, it is worthy of mention that according to the table 6, any change on the level of A factor leads to meaningful change on $(M)_{\text{Factor=I}}^{\text{Level=i}}$; therefore, the most important factor for ANN is the number of neurons in hidden layers for this data set, similar to other researches [22-24, 33].

4.3. Simulation of Structural Features

Intensities in XRD were predicted for 1 ~ 40 h milling using the OANN. These simulated patterns are depicted in figure 4. Patterns of 1, 1.5, 2, 8, 20 and 40 h are similar to the experiment samples (see figures 2 and 4). The other patterns were simulated using the OANN without experiments.

Structural features of Al_2O_3-TiB_2nanocomposite as the function of milling time are presented in the figures 2 and 4 which relates to the XRD of experiments and those one simulated by OANN. It can be concluded that increase of milling time affects the structural features. The performance accuracy of the neural network can be checked by the error of neural network simulations. For the test data set, neural network simulations are compared with the corresponding experimental values.

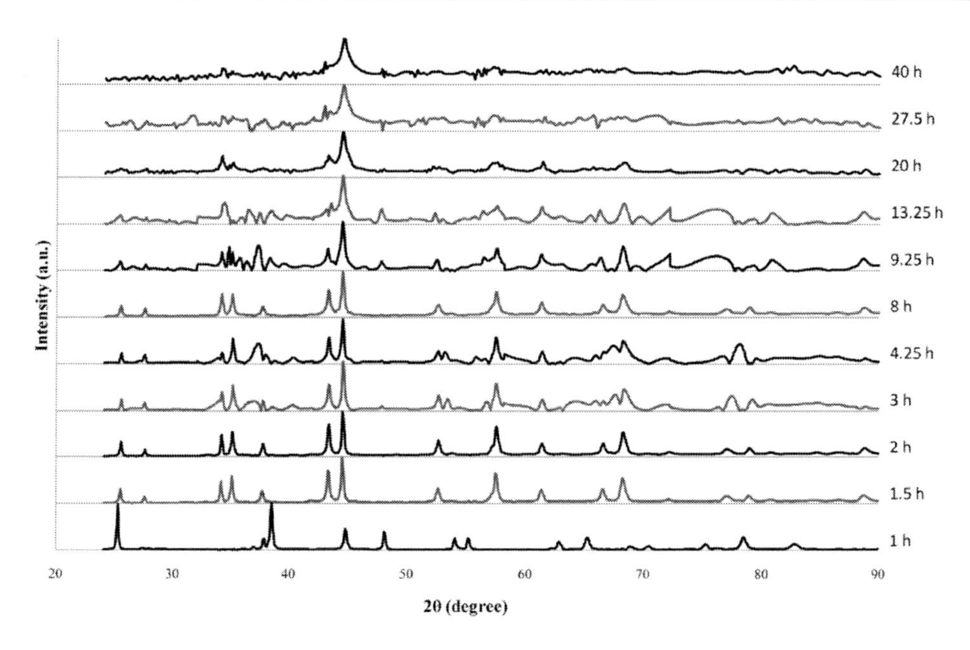

Figure 4.Simulated XRD patterns for 1 ~ 40 h milling time.

The Mean Absolute Errors (MAE) method is utilized since it is a better yardstick for small data sets. The MAE is given by:

$$MAE = \left| \frac{R_{exp.} - R_{OANN}}{R_{exp.}} \right| \tag{16}$$

in which $R_{exp.}$ are the values of experimental data and R_{OANN} are the simulated values of OANN. The amount of amorphization was calculated for both of XRD patterns of experiment and simulated samples at the same time by equation 3. After 1.5 h milling, there are two components, i.e. Al_2O_3 and TiB_2 with a difference in hardness (Al_2O_3 is harder than TiB_2) which causes difference in amorphous ratio. It should be noted that there is no formula for calculating the amorphous ratio separately. However, the amount of amorphization is the same for both of experimental and simulation data (MAE = 5.99%). The figure 5 shows that the amount of amorphization is not significant at the beginning, but it increases in a linear manner during 8-40 h of milling for both of experiment and prediction data. Based on the fact that the ANN is not valid for extrapolating [33], the results for 40 h of milling are not simulated.

Results of the interplaner distance calculation are shown in the figure 6 which they are obtained using equation 5. According to this figure, during the milling time the distance of crystalline planer is not significantly different, especially for Al_2O_3. Owing to the fact that amount of the interplaner distance is only a function of mean peak angles; MAEs are lower than 0.5%. This is due to the fact that the mean peaks of both materials are very close to each other and in the most cases they have overlap. Hence, modeling and simulation of structural properties are very tough. In both cases, TiB_2 and Al_2O_3, mean peaks for the simulated data are higher than the experimental ones leading the simulated interplaner distances to be lower than experimental results.

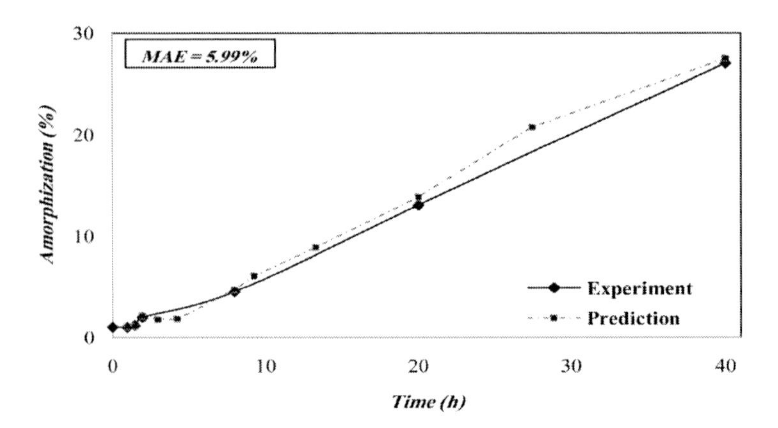

Figure 5. Comparison of the predicted (OANN) and experimental (XRD) results for the percentage of amorphization versus milling time.

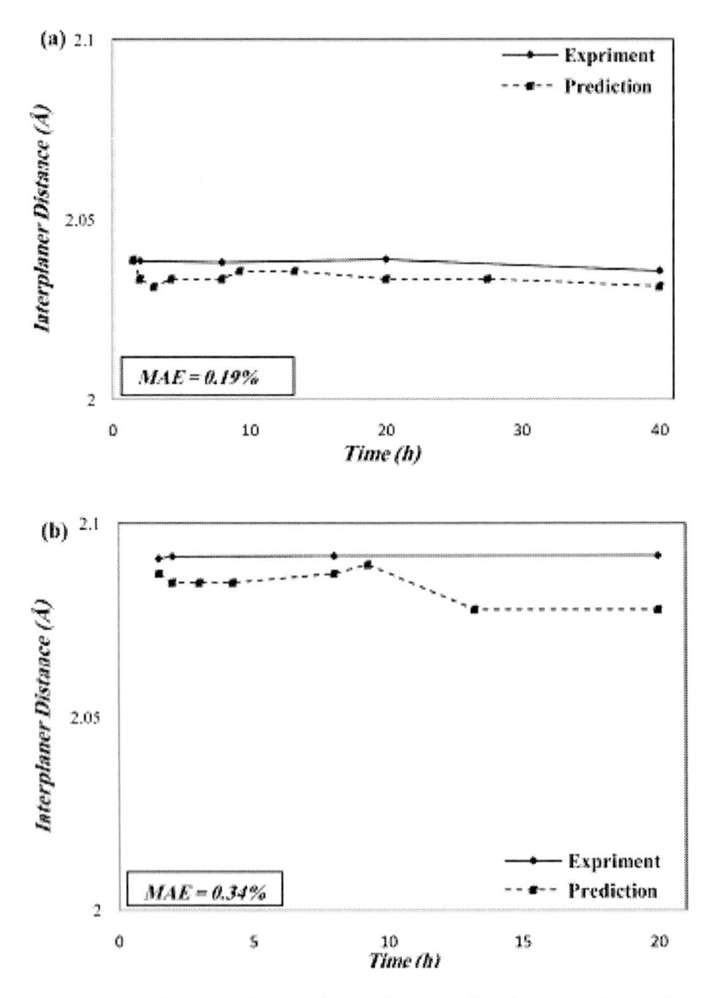

Figure 6. Comparison of the predicted and experimental results for the interplaner distance of (a) TiB_2 and (b) Al_2O_3.

Simulated values and experimental results for the crystal size and the lattice strain calculations are depicted in figures 7 and 8 using equations 4 and 6, respectively. Contrary to the interplaner distance, the crystal size and lattice strain for TiB_2 and Al_2O_3 are different. Additionally, since equations 4 and 6 are functions of mean peak angles and intensities, MAEs of the crystal size and the lattice strain are higher than MAEs of the interplaner distance.

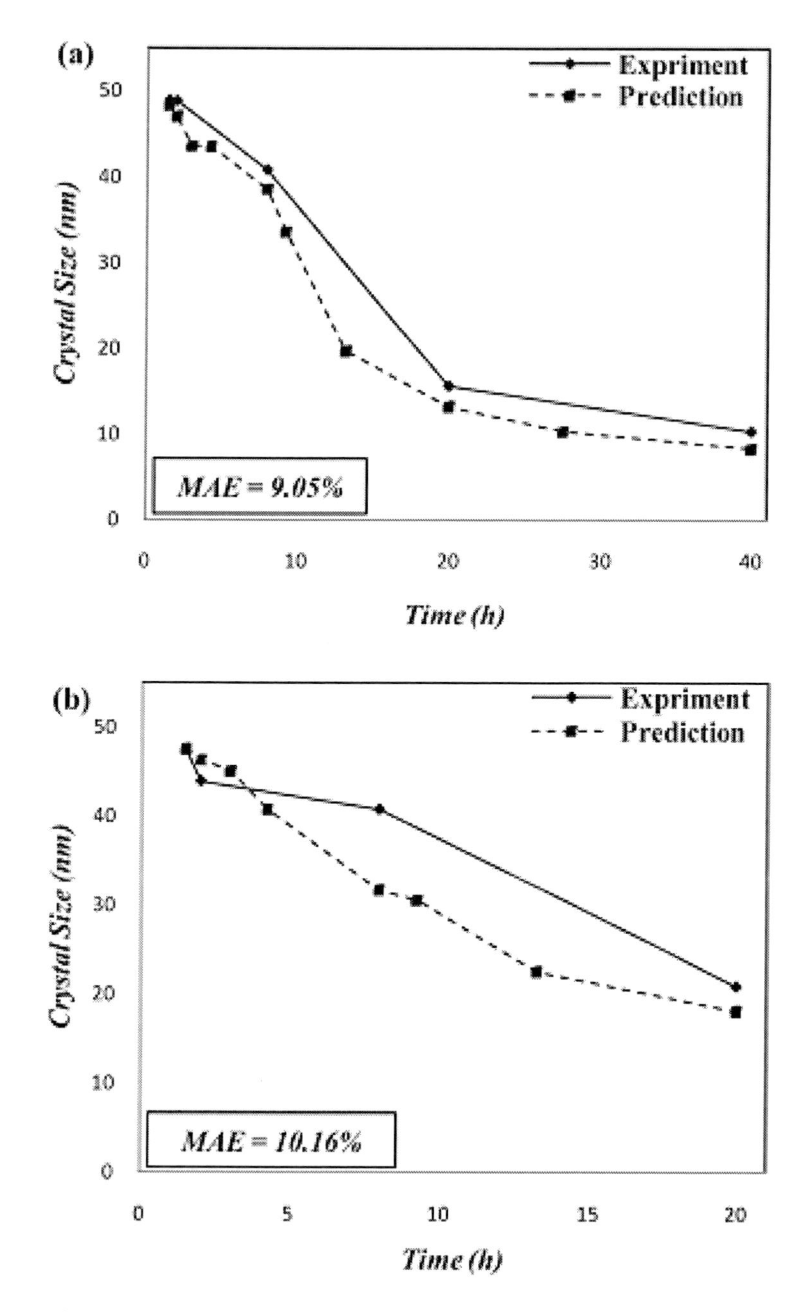

Figure 7. Comparison of the predicted and experimental results for the crystallite size of (a) TiB_2 and (b) Al_2O_3.

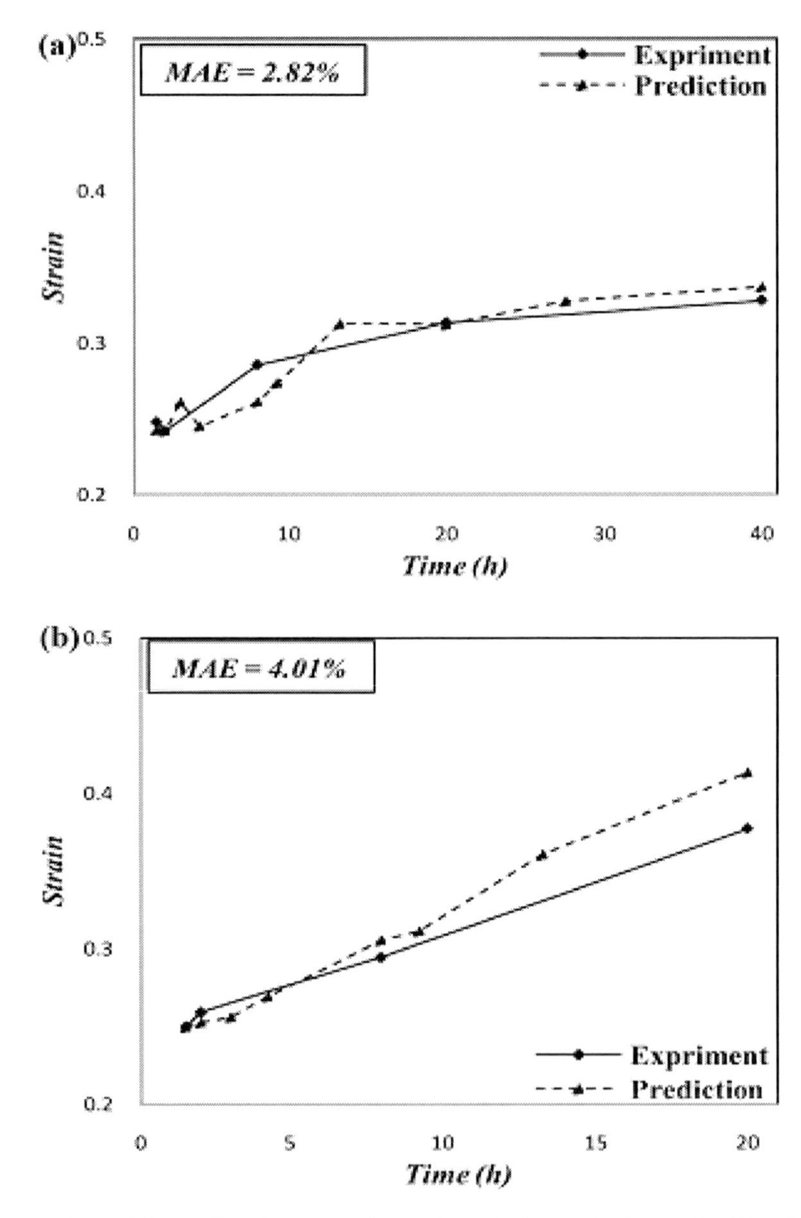

Figure 8. Comparison of the predicted and experimental results for the lattice strain of (a) TiB_2 and (b) Al_2O_3.

Conclusions

In this work, usage of AI techniques was presented for simulating the structural features on mechanochemical synthesis of Al_2O_3-TiB_2nanocomposite. Al_2O_3-TiB_2nanocomposite was synthesized from the mixtures of TiO_2, aluminum andboric acidvia a mechanochemical process. By improving the milling conditions, synthesis time decreased with high purity. XRD patterns of samples were implemented for the AI after cleaning and normalizing of data sets. More accurate results were obtained with the ANN model as compared to the

ANFISone. Besides, the ANN parameters were optimized using Taguchi method. The most important factor for ANN was the number of neurons in hidden layers. The OANN was utilized for simulating the XRD intensity in the other milling times. The crystallite size, the interplaner distance, the amorphization degree and the lattice strain were simulated using simulation results of ANN. Among the structural features, simulated values of amorphization degree and lattice strain were more acceptable in terms ofgraphical presentations and MSE values.

REFERENCES

[1] D. A. Hoke, D. K. Kim, J. C. LaSalvia and M. A. Meyers, *Journal of the American Ceramic Society*79 (1), 177-182 (1996).

[2] M. Gu, C. Huang, S. Xiao and H. Liu, *Materials Science and Engineering*: A 486 (1-2), 167-170 (2008).

[3] S. K. Mishra, S. K. Das and L. C. Pathak, *Materials Science and Engineering*: A 426 (1-2), 229-234 (2006).

[4] S. K. Mishra, S. K. Das and V. Sherbacov, *Composites Science and Technology*67 (11-12), 2447-2453 (2007).

[5] A. R. Keller and M. Zhou, *Journal of the American Ceramic Society*86 (3), 449-457 (2003).

[6] Z. Liu, S. Raynova, D. Zhang and B. Gabbitas, *Materials Science and Engineering*: A 449-451, 1107-1110 (2007).

[7] C. Suryanarayana, *Progress in Materials Science*46 (1-2), 1-184 (2001).

[8] J. M. Xue, D. M. Wan and J. Wang, *Solid State Ionics*151 (1-4), 403-412 (2002).

[9] J. Wang, J. Xue and D. Wan, *Solid State Ionics*127 (1-2), 169-175 (2000).

[10] S. Agatonovic-Kustrin, V. Wu, T. Rades, D. Saville and I. G. Tucker, *International Journal of Pharmaceutics* 184 (1), 107-114 (1999).

[11] S. Agatonovic-Kustrin, V. Wu, T. Rades, D. Saville and I. G. Tucker, *Journal of Pharmaceutical and Biomedical Analysis*22 (6), 985-992 (2000).

[12] D. Wright, C. L. Liu, D. Stanley, H. C. Chen and J. H. Fang, *Computers and Geosciences*19 (10), 1429-1443 (1993).

[13] S. Bose, D. Shome and C. K. Das, *Archives of Computational Materials Science and Surface Engineering* 1 (4), 197-204 (2009).

[14] S. Agatonovic-Kustrin, B. D. Glass, M. Mangan and J. Smithson, *International Journal of Pharmaceutics*361 (1-2), 245-250 (2008).

[15] J. S. R. Jang, presented at the *The Ieee International Conference On Fuzzy Systems*, 1996 (unpublished).

[16] R. Roy, *A Primer on the Taguchi Method.* (Van Nostrand Reinhold, New York, 1990).

[17] H. Baseri, S. M. Rabiee, F. Moztarzadeh and M. Solati-Hashjin, *Materials and Design*31 (5), 2585-2591 (2010).

[18] G. J. Myatt, *Making Sense of Data A Practical Guide to Exploratory Data Analysis and Data Mining*. (John Wiley and Sons, Inc., Hoboken, NJ, 2007).

[19] J. Jang, *Ieee Transactions On Systems, Man, And Cybernetics*23 (3), 665-685 (1993).

[20] T. T. Ajaal and R. W. Smith, *Journal of Materials Processing Technology*209 (3), 1521-1532 (2009).
[21] W. B. Lei Yingjie, *Journal of Systems Engineering and Electronics*16 (3), 583-587 (2005).
[22] H. Demuth and M. Beale, *Neural network toolbox for use with MATLAB*. (The MathWorks, Inc., Massachusetts, 1998).
[23] C. G. Zampronio, J. J. R. Rohwedder and R. J. Poppi, *Chemometrics and Intelligent Laboratory Systems* 62 (1), 17-24 (2002).
[24] V. R. Adineh, C. Aghanajafi, G. H. Dehghan and S. Jelvani, *Optics and Laser Technology*40 (8), 1000-1007 (2008).
[25] Y. Watanabe, T. Umegaki, M. Hashimoto, K. Omata and M. Yamada, *Catalysis Today*89 (4), 455-464 (2004).
[26] A. Das, J. Maiti and R. N. Banerjee, *Expert Systems with Applications*37 (2), 1075-1085 (2010).
[27] S. M. Sapuan and I. M. Mujtaba, *Composite materials technology neural network applications*. (CRC Press, Florida 2010).
[28] S. Trenn, Neural Networks, *IEEE Transactions on*19 (5), 836-844 (2008).
[29] A. R. YildIz, *Computers in Industry*60 (8), 613-620 (2009).
[30] C.-S. Chou, R.-Y. Yang, J.-H. Chen and S.-W. Chou, *Powder Technology*199264-271 (2010).
[31] E. M. Sharifi, F. Karimzadeh and M. H. Enayati, *Journal of Alloys and Compounds*502 (2), 508-512 (2010).
[32] L. C. Stephen, *Journal of Intelligent and Fuzzy Systems*4 (4), 243-256 (1996).
[33] V. Capecchi, M. Buscema, P. Contucci and B. D'Amore, *Applications of mathematics in models, artificial neural networks and arts*. (Springer, Heidelberg, Dordrecht, London, New York, 2010).

In: Artificial Intelligence
Editor: Brent M. Gordon, pp. 115-125
ISBN 978-1-61324-019-9

Chapter 5

AN ARTIFICIAL INTELLIGENCE TOOL FOR PREDICTING EMBRYOS QUALITY

***Loris Nanni*[*,1], *Alessandra Lumini*[1] *and Claudio Manna*[2]**
[1]Department of Electronic, Informatics and Systems (DEIS),
Università di Bologna, Cesena, Italy
[2]Genesis IVF Centre, 00189 Rome

ABSTRACT

One of the most relevant aspects in Assisted Reproductive Technologies is the characterization of the embryos to transfer in a patient. Objective assessment of embryo quality is actually an important matter of investigation both for bioethical and economical reasons. In most cases, embryologists evaluate embryos by visual examination and their evaluation is totally subjective. Recently, due to the rapid growth in our capacity to extract texture descriptors from a given image, a growing interest has been shown on the study of artificial intelligence methods to improve success rates of IVF programs based on the analysis and selection of images of embryos or oocytes.

In this work we concentrate our efforts on the automatic classification of the quality of an embryo starting from the analysis of its image. The artificial intelligence system proposed in this work is based on textural descriptors (i.e. features), used to characterize the embryos, by measuring the homogeneity of their texture and the presence of recurrent patterns. A general purpose classifier is trained using visual descriptors to score the embryo images.

The proposed system is tested on a datasets of 257 images with valuable classification results.

Keywords: embryo selection; assisted reproduction technologies; machine learning techniques; support vector machine.

[*] E-mail: loris.nanni@unibo.it

1. INTRODUCTION

Since the introduction of in vitro fertilization (IVF) for treatment of infertility, more than 30 years ago, one of the most studied aspects of assisted reproductive technology (ART) has been the evaluation of embryo quality. In fact, the ability to evaluate the quality of the embryos in a non invasive way would help the embryologists to a better choice of embryos to be transferred.

An accurate analysis of embryos, usually on the second or third day after fertilization, in order to assess their quality is useful to select of the most potent embryos for transfer and subsequently to improve success rates of IVF programs. Moreover the selection of only the best embryo for transfer reduces the possibility of multiple pregnancy rates.

Unfortunately there is no universal consensus about general rules to determine embryo quality and embryos are selected based on different scoring system at every in vitro fertilization (IVF) centre. Moreover, the methods of embryo examination are rapidly changing during the last 20 years, also thanks to new technologies for artificial image analysis. Non-invasive embryo examination is based on observation of morphology and dynamics of embryo development, usually performed under contrast-phase microscope with Hoffmann modulation contrast (HMC) or difference-interference contrast (DIC).

As stated above, many different classification criteria have been proposed in a literature (the interested reader can refer to [34] for a survey), most of which indicate the following parameters as the ones that influence most often the selection of good quality embryos [28][29]: pronuclear morphology, polar body structure and placement, appearance of cytoplasm (pitting, vacuoles and halo effects) and zona pellucid, early cleavage, number of blastomeres in particular days of culture, size, symmetry and fragmentation of blastomeres, compaction and expansion of blastomeres, multinucleation.

Thank to these criteria, often combined in complex ways, the prediction of developmental potential of particular embryo is made more feasible thus increasing the potentiality of giving a chance for pregnancy in infertile couples. Many different systems have been proposed, each based on the individual laboratory's familiarity and training: some of them pay attention to the embryo cleavage state [30][31] while others incorporate multiple morphological criteria [32].Anyway there is no possibility of make a fair comparison among them, due to the different testing conditions.

Many works have also been published regarding other screening parameters, such as assessment of the oocyte, pronuclear as well as early cleavage status. Among these alternative scoring systems valuable results have been reported by [33] which present a computer program created to analyze zygote images semi-automatically, providing precise morphological measurements. The experimental validation has been performed on a large dataset of 206 zygotes from two different IVF centres; the dataset includes only zygotes with certain labels (84 originated from single or twin pregnancies after the transfer of one or two embryos, respectively, 122 not implanted). The classification error achieved with the computer-assisted measurements was only slightly inferior to that of the subjective ones.

In this work we concentrate our efforts on the automatic classification of the quality of an embryo starting from the analysis of its image. The artificial intelligence system proposed in this work is based on textural descriptors (i.e. features), used to characterize the embryos, by measuring the homogeneity of their texture and the presence of recurrent patterns. A general purpose classifier is trained using visual descriptors to score the embryo images. The ground truth used to train the classifier has been obtained asking to an expert embryologist to classify an embryo according to well known scoring criteria [34]; the embryo scoring is performed by visual examination of the embryo therefore having more information than doing it on a 2D image.

The proposed system is tested on a datasets of 257 images with valuable classification results.

2. Proposed System

In this work we propose a pattern recognition system which combines good texture descriptors with high performance general purpose classifiers for scoring the embryos without the work of a trained biologist.

The architecture of the system is schematized in figure 1, while a deep description of each step (pre-processing, feature extraction and classification) is given in the following sub-sections.

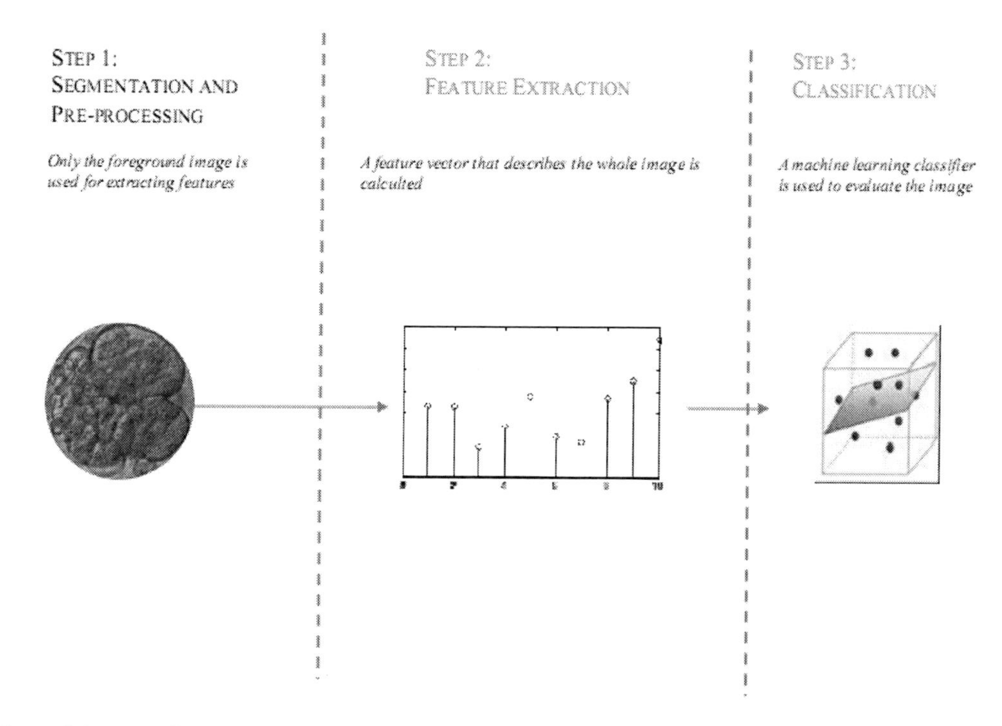

Figure 1. Proposed system for embryos image scoring.

2.1. Segmentation and Pre-Processing

The segmentation of the region of interest from the background is performed according to the active contour snake approach [12]. Snake is an elastic continuous curve that from an initial position begins to deform to adjust the object's contour. So the energy-minimizing spline is guided by external constraint forces and influenced by image force that pull it toward features such as lines and edges. External forces attract the snake towards image features. Internal forces avoid discontinuities in the snake shape.

After the segmentation, the images are rescaled to the same dimensions (100×100 pixels) and an enhancement method is performed to deal with the inherent non-uniform illumination problems (figure 2). The pre-processing method is based on a local normalization [27]: each pixel P(x, y) is normalized by using the operation below:

$$\mathbf{P}'(x,y)=\begin{cases} m_t+\beta \text{ if } \mathbf{P}(x,y)>m \\ m_t-\beta \text{ otherwise} \end{cases} \text{ where } \beta=\sqrt{\frac{v_t(\mathbf{P}(x,y)-m)^2}{v}}$$

where m and v are the image mean and variance, respectively, m_t and v_t are the parameters (both set to 100 in the experiments) for mean and variance for the output image.

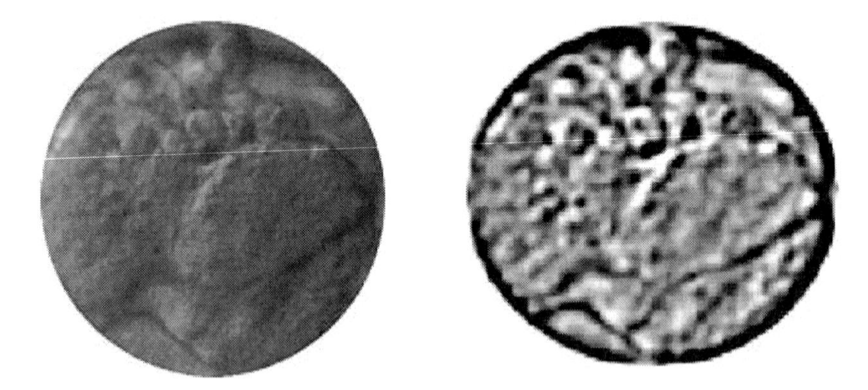

Figure 2. Segmented image of embryo (left) and its enhanced version (right).

2.2. Feature Extraction

In this work we test three different texture descriptors: Local Phase Quantization (LPQ) [24] and dominant local ternary patterns (DLTP), a novel variant of dominant local binary pattern [23], and local ternary patterns (LTP) [21].

The Local Phase Quantization (LPQ) operator was originally proposed by Ojansivu and Heikkila as a texture descriptor [24]. LPQ is based on the blur invariance property of the Fourier phase spectrum. It uses the local phase information extracted using the 2-D short-term Fourier transform (STFT) computed over a rectangular neighborhood at each pixel position of the image. In LPQ only four complex coefficients are considered, corresponding to 2-D frequencies. For more mathematical details, refer to [24]. In our experiments, we use three LPQ histogram[1] obtained fixing the radius r of the windows for the local phase information extraction to $r = \{1, 3, 5\}$.

The dominant local ternary patterns (DLTP) descriptor is a combination between the dominant local binary pattern and local ternary patterns.

1 Calculated using original MATLAB code shared by Ojansivu and Heikkila.

Dominant local binary pattern was proposed in [23] as a variant of LBP that used the patterns that represents the 80% of the whole pattern occurrences in the training data as rotation invariant patterns to be selected, instead of uniform patterns. In this work we set 250 the number of patterns that have higher variance among the training data to be selected. Our variation consisted in substitute the LBP operator with the more stable LTP operator [21].

The LBP operator is calculated by evaluating the binary differences between the gray value of a pixel x and the gray values of *P* neighboring pixels on a circle of radius *R* around x (see figure 3). In local ternary patterns (LTP) [21] the difference between a pixel x and its neighbor u is encoded by 3 values according to a threshold τ: 1 if $u \geq x + \tau$; -1 if $u \leq x - \tau$; else 0. The ternary pattern is then split into two binary patterns by considering its positive and negative components. Finally, the histograms that are computed from the binary patterns are concatenated to form the feature vector. In our implementation only uniform patterns, i.e. it contains at most two bitwise transitions from 0 to 1 or vice versa, are used to construct the histogram.

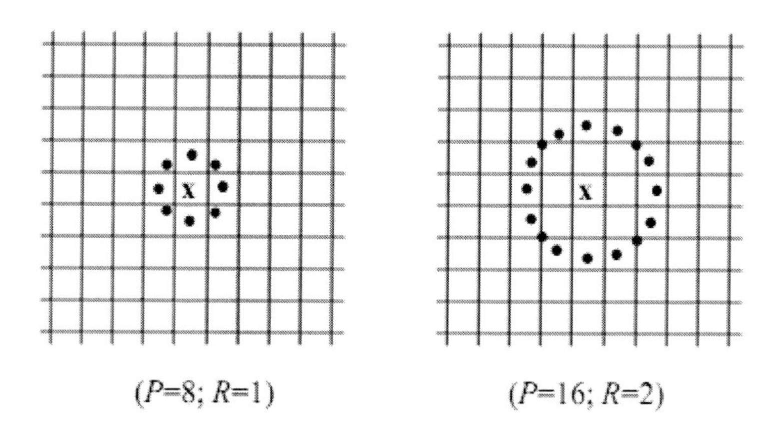

Figure 3. LBP/LTP neighbor sets for (*P*=8; *R*=1) and (*P*=16; *R*=2).

2.3. Classification

The classification step has been performed by training a support vector machine[2] (SVM) for each descriptor. The support vector machine is a technique for classification from the field of statistical learning theory [22]. SVM is a binary-class prediction method trained to find the equation of a hyperplane that divides the training set leaving all the points of the same class on the same side while maximizing the distance between the two classes and the hyperplane. In cases where a linear decision boundary does not exist a kernel function can be used: a kernel function allows to project the data onto a higher-dimensional feature space in which it may be separable by a hyperplane. Typical kernels are polynomial kernels and radial basis function kernel.

Notice that, before the classification, all the features used for training SVM are linearly normalized to [0 1] considering the training data.

The final score has been obtained by two ensembles of SVM classifiers:

2 SVM is implemented as in the OSU svm toolbox.

- The three LPQ histogram obtained varying the radius *r* are concatenated to form the feature vector used to train the SVM classifier;
- The DLTP descriptors are used to train a random subspace (RS) ensemble of SVMs (since it is reported in the literature [26] that with DLTP the RS method permits to improve the performance).
- The LTP descriptors are used to train a random subspace (RS) ensemble of SVMs.

Random subspace [6] is a method for reducing dimensionality by randomly sampling subsets of features (50% of all the features in our experiments). RS modifies the training data set by generating K (K=*50* in our experiments) new training sets. It builds classifiers on these modified training sets. In other words, each classifier is trained on each of the new training sets. The results are combined using the sum rule.

3. Embryos Dataset

The experiments have been carried out on a dataset of 257 photographs of embryos usually at the 4-cell stage taken 40–50 hours after fertilization and before transfer to uteri. The photographs have been taken by placing each embryo under a microscope (Inverted Microscope Olympus IX 70; Olympus America Inc., Melville, NY, USA), with a camera (video camera JVC TK-C401EG; Victor Company of Japan, Limited (JVC), Yokohama, Japan). Examples of the images used in the experimentation can be viewed in figure 4.

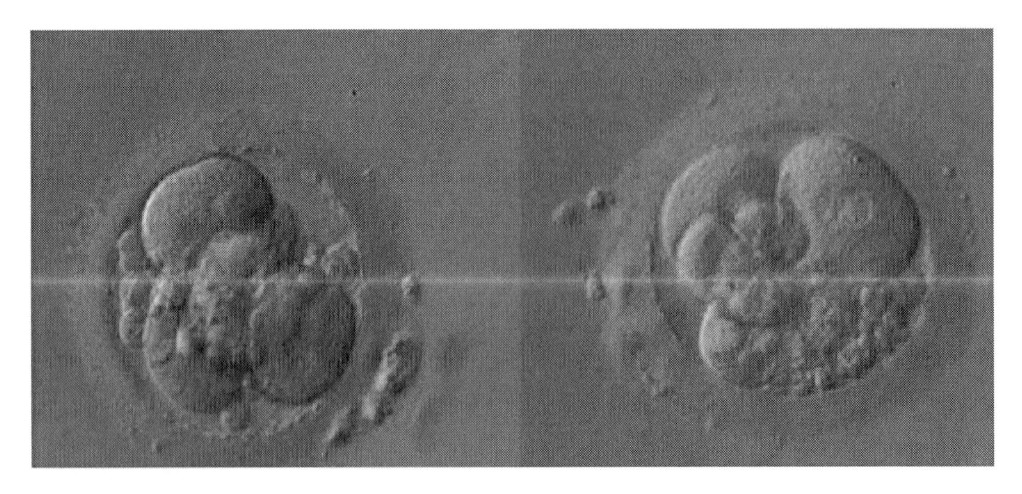

Figure 4. Some samples from the dataset.

The research was conducted at a single fertility centre and involved 94 women, with different infertility factors (unexplained, tubal, male), in a total number of 94 transfer cycles with transfer of intracytoplasmic sperm injection (ICSI).

Ovarian stimulation was carried out by administering recombinant FSH (Gonal F; Merck Serono S.A., Geneva, Switzerland) at a dosage of 150–400 IU according to individual response after suppression with gonadotrophin-releasing hormone analogue in a daily preparation (Suprefact; Hoechst Marion Roussel Deutschland, Bad Soden Germany). Oocyte retrieval was carried out with ultrasound-guided transvaginal follicular aspiration after 35 hours from the administration of 5000 IU of human chorionic gonadotrophins (Profasi, Serono). Gametes and embryos were cultured under oil in drops of a culture medium (IVF Scandinavia; Vitrolife Sweden AB, Kungsbacka, Sweden) with an atmosphere of CO_2 of 5% in air. The ICSI process was performed according to current methodology [20].

According to the quality classification used in [34], in this work we have used two different quality scoring for embryos (some samples are shown in figure 5):

- the first (named "Scoring A vs B") is based on presence of early first cleavage and on the number of the individual cells (referred to as blastomeres). The good quality embryo has at least 4 cells on the second day and at least 8 cells on the third day of culture while four or less blastomeres on the third day of culture indicates low developmental potential.
- the latter (named "Scoring 1 vs 2-3-4") is based on degree of regularity of size of blastomeres and degree of fragmentation. A good quality embryo is without or with negligible fragmentation, the embryos are categorized considering their level of fragmentation. Embryos with more than 25% fragmentation have a low implantation rate.

The "true" class of each embryo is assigned by an expert biologist by visually inspection the embryo at the microscope. In our dataset, table 1, the combination between the two quality scoring is the best method for choosing the embryos. These results are calculated considering the cycles where all the embryos of a given woman belong to the same quality class. The row ALL reports the results obtained considering all the embryos of the dataset.

Unfortunately the small dimension of the dataset does not permit to confirm, by a statistical analysis, if the differences among the pregnancies rate (i.e. the percentage of all attempts that lead to pregnancy) of the different groups are not random.

Table 1. Implantation/ Pregnancies rate in function of the quality score

Quality	Number of cycles	Number of embryos	Pregnancies Rate	Implantation Rate
A	46	124	45.65%	22.58%
B	23	50	26.09%	18.00%
1	57	155	43.86%	22.58%
2-3-4	7	12	14.29%	16.67%
1A	28	78	60.71%	29.49%
1B	14	28	14.29%	10.71%
2A-3A-4A	3	7	00.00%	00.00%
2B-3B-4B	4	5	25.00%	40.00%
ALL	94	257	39.6%	19.84%

In order to compare our automatic scoring system to human experts, two skilled embryologists were asked to give a quality score to each embryo, analyzing its 2D image, according to the same scoring protocol using to label the dataset (a first quality score with two values A or B, and a second with four values 1, 2, 3 and 4).

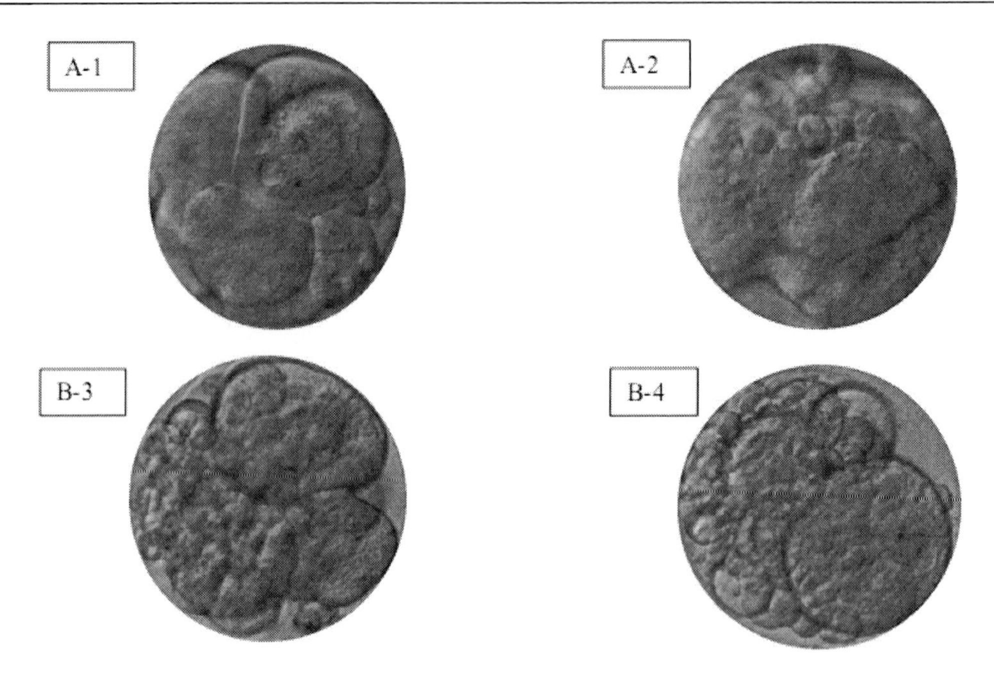

Figure 5. Samples of scored images.

4. RESULTS

The aim of this section is to evaluate the proposed approach on the available dataset, according to a fold cross validation testing protocol. We performed 100 experiments obtained randomly dividing the images between training and testing set. In each experiment the images related to the 10% of women are randomly selected to be removed from the training and used as test set; the average results on the 100 experiments are reported in the following tables.

As performance indicator we have used the area under the ROC curve (AUC) [4]; this performance indicator can be interpreted as the probability that the classifier will assign a lower score to a randomly picked positive sample than to a randomly picked negative sample. It can also be shown that the area under the ROC curve is closely related to the Mann–Whitney U test, which tests whether positives samples are ranked higher than negatives (while the old error rate indicator does not consider the scores of the classifiers).

The first test is aimed at comparing different texture descriptors for the classification of the quality of an embryo starting from its image.

The following texture descriptors are here compared:

- DLTP, Dominant Local Ternary Patterns, as described in section 2 (no image normalization is performed before this feature extraction);
- LTP, Local Ternary Patterns (no image normalization);
- LPQ, Local Phase Quantization (image normalization is performed before extracting this descriptor);
- SUM, combination by sum rule of the results of DLTP, LTP and LPQ. Notice that before the fusion the scores have been normalized between [0,1].

Table 2. AUC obtained on the Embryo dataset

	DLTP	LTP-u	LPQ	SUM	Embryologist 1	Embryologist 2
Scoring 1 vs 2-3-4	0.78	0.78	0.82	0.84	0.88	0.86
Scoring 1A vs other	0.72	0.74	0.76	0.78	---	---

It is clear that the texture descriptors permit to obtained a reliable result. The two trained biologists obtains only slight better results in the "Scoring 1 vs 2-3-4" problem when they assign to the embryo a quality scores from its image. We want to stress that in the proposed approach also the image segmentation is automatically performed. In our opinion a larger dataset will permit to obtain performance similar to those obtained by humans. For example using a subset of 146 embryos the method DLTP obtains an AUC of 0.68 while LPQ obtains an AUC of 0.79. It is clear the performance improvement when all the 257 embryos are used (0.68->0.78 for DLTP, 0.79->0.82 for LPQ).

Another interesting result is that LPQ is the best approach and the fusion among the methods outperforms LPQ also when the simple sum fusion rule is used.

Discussion

One of the aims of Assisted Reproduction Technologies in this era is to improve the pregnancy rate minimizing the probability of multiple pregnancies, since the multiple pregnancies drastically increase the rate of preterm deliveries, the risk of lifelong disability and often compromises the survival of neonates [1].

In this work we propose an automated system for the textural analysis of the embryos images, our system analyzes and scores digital images of embryos based on a textural descriptor.

The best practical findings revealed in this work is that it is possible to obtain an enough reliable embryo scoring starting from its image without the help of trained physicians or biologists. In our opinion when large datasets will be available and more suited texture descriptors will be developed/tested will be possible to train a machine that obtains the same performance of a trained human.

References

[1] Botros L., Sakkas D. and Seli E., Metabolomics and its application for non-invasive embryo assessment, in *Molecular Human Reproduction* 2008 14(12):679-690.

[2] Duda R., Hart P., Stork D., (2001) *Pattern Classification*, Wiley, New York.

[3] Elfadel I.M. and Picard R.W., "Gibbs random fields, cooccurrences, and texture modeling," *IEEE Trans. On PAMI*, vol. 16, no. 1, pp. 24–37, 1994.

[4] Fawcett T. (2004). "ROC Graphs: Notes and Practical Considerations for Researchers". *Technical report, Palo Alto*, USA: HP Laboratories.

[5] Hagan M.T., and Menhaj M. (1994). Training feed-forward networks with the Marquardt algorithm, *IEEE Transactions on neural networks*, 5(6), 989-993.

[6] Ho TK. (1998) The random subspace method for constructing decision forests, *IEEE Transactions on Pattern Analysis and Machine Intelligence*, 20 (8) 832–844.

[7] Jurisica I., Mylopoulos J., Glasgow J., Shapiro H., Casper R., Case-based reasoning in IVF: prediction and knowledge mining, *Artif. Intel. Med.* 12 (1998) 1–24.

[8] Kruizinga P., Petkov N., and Grigorescu S.E., "Comparison of texture features based on gabor filters," in *Proc. of the 10th ICIAP*, 1999, p. 142.

[9] Ma W.Y. and Manjunath B.S., "A comparison of wavelet transform features for texture image annotation," in *Proc.of the ICIP*, 1995, p. 2256.

[10] Manna C., Patrizi G., Rahman A., Sallam H., Experimental results on the recognition of embryos in human assisted reproduction, *Reprod. BioMed. Online* 8 (2004) 460–469.

[11] Morales D. A., Bengoetxea E. and Larrañaga P. 2008. Selection of human embryos for transfer by Bayesian classifiers. *Comput. Biol. Med.* 38, 11-12 (Nov. 2008), 1177-1186.

[12] Morales D., Bengoetxea E., Larrañaga P. (2008). Automatic Segmentation of Zona Pellucida in Human Embryo Images Applying an Active Contour Model. Proceedings of the Medical Image Understanding and Analysis (MIUA 2008), 209-213. J. Hoey and S. McKenna (eds.). Dundee, Scotland.

[13] Morales D.A., Bengoetxea E., Larrañaga P., García M., Franco Y., Fresnada M. and Merino M. 2008. Bayesian classification for the selection of in vitro human embryos using morphological and clinical data. *Comput. Methods Prog. Biomed.* 90, 2 (May. 2008), 104-116.

[14] Nanni L. and Lumini A. (2008b), A reliable method for cell phenotype image classification, *Artificial Intelligence in Medicine*, 43 (2) Pages 87-97.

[15] Ojala T, Pietikainen M, and Maeenpaa T. (2002) Multiresolution Gray-Scale and Rotation Invariant Texture Classification with Local Binary Patterns, *IEEE Transactions on Pattern Analysis and Machine Intelligence*, 24 (7) 971-987.

[16] Patrizi G., Manna C., Moscatelli C., Nieddu L., Pattern recognition methods in human-assisted reproduction, *Int. Trans. Oper. Res.* 11 (2004) 365–379.

[17] Pizer S.M., Amburn E.P., Austin J.D., et al. Adaptive Histogram equalization and its variations. *Computer Vision, Graphics, and Image Processing*, 1987, 39(3): 355~368.

[18] Saith R., Srinivasan A., Michie D., Sargent I., Relationships between the developmental potential of human in-vitro fertilization embryos and features describing the embryo, oocyte and follicle, *Human Reprod.* Update 4 (1998) 121–134.

[19] Trimarchi J.R., Goodside J., Passmore L., Silberstein T., Hamel L., Gonzalez L., Comparing data mining and logistic regression for predicting IVF outcome, *Fertil. Steril.* 80 (2003), 100–100.

[20] Van Steirteghem AC, Nagy Z, Joris H, et al. High fertilization and implantation rates after intracytoplasmic sperm injection. *Hum. Reprod.* 1993;8:1061–6.

[21] X. Tan, B. Triggs, "Enhanced Local Texture Feature Sets for Face Recognition under Difficult Lighting Conditions", Proc. *IEEE International Workshop on Analysis and Modeling of Faces and Gestures,* pp. 168-182, 2007.

[22] Cristianini N, Shawe-Taylor J (2000). *An introduction to Support vector machines and other kernel-based learning methods,* Cambridge University Press.

[23] Liao S., Law M.W.K. and Chung A.C.S. (2009) Dominant Local Binary Patterns for Texture Classification, *IEEE Transactions on Image Processing*, (TIP, 2009), Vol. 18, No. 5, pages 1107 – 1118, May, 2009.

[24] Ojansivu V. and Heikkila J. (2008) Blur insensitive texture classification using local phase quantization. In *ICISP* 2008.

[25] Ahonen T, Matas J, He C and Pietikäinen M (2009), Rotation invariant image description with local binary pattern histogram fourier features, Image Analysis, SCIA 2009 Proceedings, *Lecture Notes in Computer Science* 5575, 61-70.

[26] Loris Nanni, Sheryl Brahnam, Alessandra Lumini. 'A study for selecting the best performing rotation invariant patterns in local binary/ternary patterns.' In *Proceedings of the International Conference on ImageProcessing, Computer Vision, and Pattern Recognition* (IPCV'10), 2010.

[27] T. Connie, A.T.B. Jin, and M.G.K. Ong, D.N.C. Ling, "An automated palmprint recognition system", *Image and Vision Computing* 23 (2005) 501–515.

[28] Van Royen E, Mangelschots K, De Neubourg D, et al. Characterization of a top quality embryo, a step towards single-embryo transfer. *Hum. Reprod.* 1999; 14:2345- 9.

[29] Van Royen E, Mangelschots K, Vercruyssen M, et al. Multinucleation in cleavage stage embryos. *Hum. Reprod.* 2003; 18:1062-9.

[30] Ziebe S, Petersen K, Lindenberg S, et al. Embryo morphology or cleavage stage: how to select the best embryos for transfer after in-vitro fertilization. *Hum. Reprod.* 1997; 12:1545-9.

[31] Saith RR, Srinivasan A, Michie D, Sargent IL. Relationships between the developmental potential of human in-vitro fertilization embryos and features describing the embryo, oocyte and follicle. *Hum. Reprod. Update* 1998; 4:121-34.

[32] Desai NN, Goldstein J, Rowland DY, Goldfarb JM. Morphological evaluation of human embryos and derivation of an embryo quality scoring system specific for day 3 embryos: a preliminary study. *Hum. Reprod.* 2000; 15:2190-6.

[33] Beuchat A, Thévenaz P, Unser M, Ebner T, Senn A, Urner F, Germond M, Sorzano CO, Quantitative morphometrical characterization of human pronuclear zygotes. *Hum. Reprod.* 2008 Sep;23(9):1983-92. Epub 2008 Jun 7.

[34] Baczkowski T, Kurzawa R, Glabowski W. Methods of embryo scoring in in vitro fertilization. *Reprod. Biol.* 2004;4:5–22.

In: Artificial Intelligence
Editor: Brent M. Gordon, pp. 127-141
ISBN 978-1-61324-019-9

Chapter 6

PASSIVE SYSTEM RELIABILITY OF THE NUCLEAR POWER PLANTS (NPPS) USING FUZZY SET THEORY IN ARTIFICIAL INTELLIGENCE

Tae-Ho Woo*

Department of Nuclear Engineering, Seoul National University,
Seoul, Republic of Korea

ABSTRACT

The new kind of probabilistic safety assessment (PSA) method has been studied for the very high temperature reactor (VHTR) which is a type of nuclear power plants (NPPs). There is a difficulty to make the quantification of the PSA, because the operation and experience data are deficient. Hence, it is necessary to manipulate the data statistically in basic events. The non-linear fuzzy set algorithm is used to quantification of the designed case for the physical data. The mass flow rate of the natural circulation is a main model. In addition, the potential energy in the gravity, the temperature and pressure in the heat conduction, and the heat transfer rate in the internal stored energy are also investigated. The values in the probability set and the fuzzy set are compared for the failure analysis. The results show the failure frequency in the propagations. It is concluded the artificial intelligence analysis of the fuzzy set could enhance the reliability than the probabilistic analysis.

Namely, the dynamical safety assessment for the rare events has been newly developed. It is analyzed for the non-linear algorithm to substitute the probabilistic descriptions. The characteristics of the probabilistic distribution like the mean value and the standard deviation are changed to the some geometrical configuration as the slope and the radius in the fuzzy distribution. So, the operator can express the conceptions of the physical variable much more easily and exactly. The meaning of the non-linear algorithm shows the priority of the analysis.

Using interpretation of the fuzzy set distributions, the quantity of the physical variables can be showed with the linguistic expression of the operator. Therefore, for the further study, the human error could be reduced due to the human oriented algorithm of

* Department of Nuclear Engineering, Seoul National University, Gwanak 599, Gwanak-ro, Gwanak-gu, Seoul, 151-742, Republic of Korea. Tel. : 82-2-880-8337. Fax : 82-2-889-2688. E-mail : thw@snu.ac.kr

the theory in some active systems, because the fuzzy set theory is originated from the linguistic expression of the operator. In addition, the other complex algorithm like the neural network or chaos theory could be applied to the data quantification in PSA.

Keywords: Fuzzy Set Theory, Nuclear Power Plants, Very High Temperature Reactor, Passive System, Anticipated Transient Without Scram, Statistical Method.

1. INTRODUCTION

In the nuclear industry, one of important issue is to construct the gas-cooled nuclear power plant for the commercial purpose. It, however, is very difficult for the operator to make the quantification of the safety assessment in the uncommon event. There is no generic data in the safety analysis for the natural circulation of the reactor. Hence, it is necessary to make the advanced technology to obtain the data of the basic event. The passive system is originated from the natural circulation of the coolant. This is done due to the decay heat removal in the nuclear power plants (NPPs). In the very high temperature reactor (VHTR) of the gas-cooled reactor (GCR), the coolant is a gas type, which is the one phase in the normal operation coolant. However, in the commercialized pressurized water reactor (PWR) of the light water reactor (LWR), the phase could be changed from liquid to gas. So, the safety aspect shows a difference in the decay heat removal from the reactor. The key point of the safety feature in VHTR is the fuel temperature, which affects to the sequence of an accident. The computational quantification of the physical value is very important before the experiment and the operation have been done. The statistical investigation is useful to find out the safety assessment in the interested event.

The impact-affordability algorithm depends on the safety margin. The impact is the load of an event and the affordability is the capacity of an event. This concept is applied from a previous report (Bianchi, 2001), which is focusing on the functional failure. The normal distribution is used to make the safety margin in the event. The difference between two distributions of the event could make the interpretations of the several variables in the system like the gravity and internal force. The natural circulation in the decay heat removal of the VHTR's anticipated transient without scram (ATWS) is investigated for the passive system using the impact-affordability algorithm. This event is the model in the design basis accident (DBA).

The main safety concept for the DBA in some new generation NPPs is the passive system, because this can enhance the reliability. The gas-turbine-modular helium reactor (GT-MHR) of the General Atomics is used for the modeling. The DBA was constructed by the Korea Atomic Energy Research Institute (KAERI), which is based mainly on the license procedures of the Fort Saint Vrain (FSV) Reactor (KAERI, 2007). The section 2 explains the method of the study. The calculation for the modeling is shown in the section 3. The section 4 describes results of the study.

2. METHOD

The ATWS in VHTR is analyzed using the impact-affordability algorithm. The failure distribution of an event is described by the impact. The Figure 1 shows the relationship between the impact and affordability in the case of the normal distribution and the fuzzy distribution (Triangular case). The difference between two graphs in each case shows the safety margin. The graphs can show the distributions of mass flow rate, potential energy, temperature, pressure, and hear transfer rate in passive system. Figure 2 shows the probabilities of failure for several variables (Nayak, 2007), which are modified from the case of the mass flow rate. Using the newly made quantity of Figure 2, the probability of failure for some other basic accidents are decided, where the operational data are also deficient. For the constructing the probability of failure, it is necessary to decide the standard. The mass flow rate is the standard variable which makes the probability of failure. The other probabilities of failure in the non-operational data can be made by the proportional value to the probability of failure made by the mass flow rate. In this study, the mass flow rate of the long-term cooling in the natural circulation is used. For the other application, the other probabilities of failure could be used. There is the simplified natural circulation in Figure 3. Table 1 shows the event and scenario of the ATWS of this paper. This describes the event and scenario of the long-term cooling where the natural circulation is performed. The flow-chart of the analysis step is made in Table 2.

Table 1. Event and scenario of ATWS

Event	Scenario
Initiating event	Rod withdrawal and scram failure
Pre-turbine trip	SCS(shutdown cooling system) fails to start
Flow coastdown and power equilibrium	Core power increases (Local ϕ_n increases) by reactivity insertion
	T_{core} and P_{sys} increase
	Power control by runback
	Reactor trip signal $\rightarrow$ turbine trip, but, no reactor scram
	Coastdown of primary flow
	PCM to rapidly heatup core
	Core power decreases by doppler feedback, Xe inventory increases
	Equilibrium power level $\approx$ Decay heat
Recriticality and long-term cooling	Long-term conduction and radiation cooling
	Re-criticality and power oscillation

3. CALCULATION

For the quantification of the PSA, there are two kinds of methods which are using liner and non-linear statistical data. The linear statistical data is obtained by the mass flow rate and its related variables. The statistical data are used instead of the real operation data due to the shortage of the real data. Each distribution is considered as the normal distribution (Burgazzi,

2007). The mass flow rate is used to find the safety margin for the physical values. The probability of the failure of the mass flow rate is found by the assumed distribution in Figure 2. The safety margin as well as the failure frequencies in other physical variables is assumed to be changed linearly. For the non-linear complex logic, the fuzzy set theory is used (Zio, 2003), which is modified in this paper for much more reasonable analysis. Using the definition of safety margin as sm(x) = I(x) – A(x) (I(x) : Impact function, A(x) : Affordability function), one can find as follows,

sm(x) > 0 for safe functions

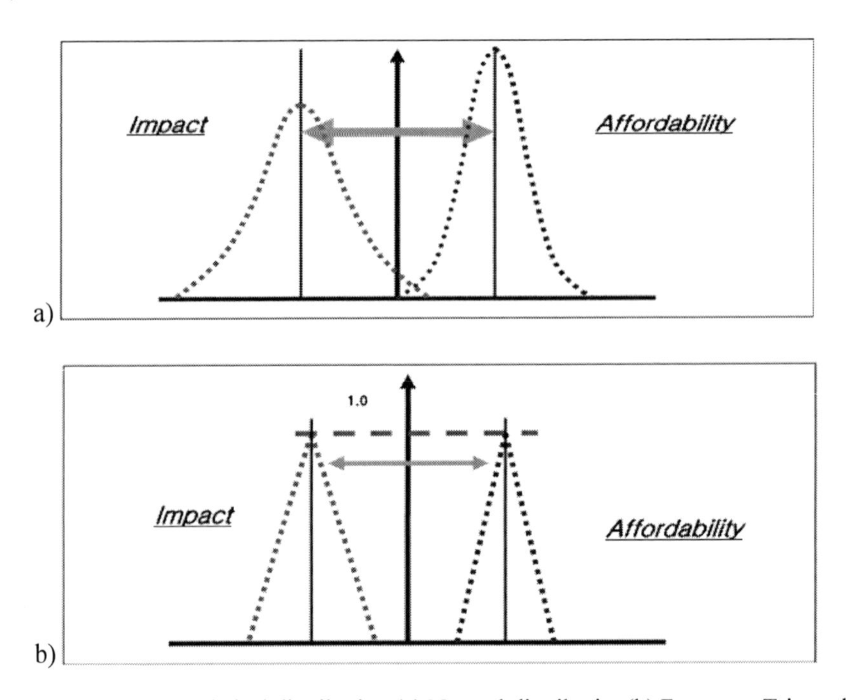

Figure 1. I-A algorithm by statistical distribution (a) Normal distribution(b) Fuzzy set-Triangular.

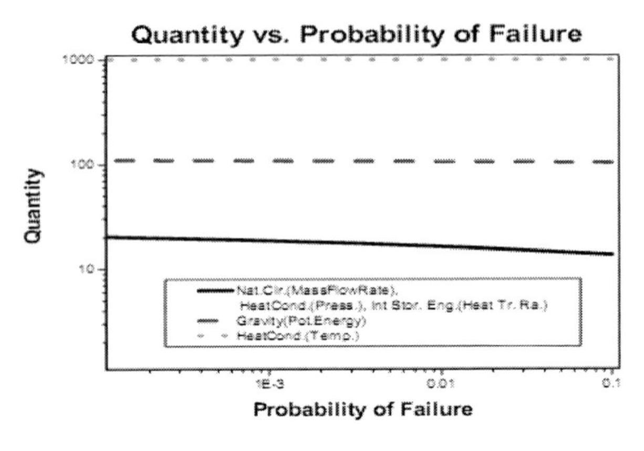

Figure 2. Comparison between probability of failure and quantity: Natural circulation (Mass flow rate, $\mu_A = 10 kg/s$), Gravity (Potential energy, $\mu_A = 100 Mw/hr$), Heat conduction (Temperature, $\mu_A = 1000\ ^oC$), Heat conduction (Pressure, $\mu_A = 10 MPa$), Internal stored energy (Heat transfer rate, $\mu_A = 10 kw/s$)

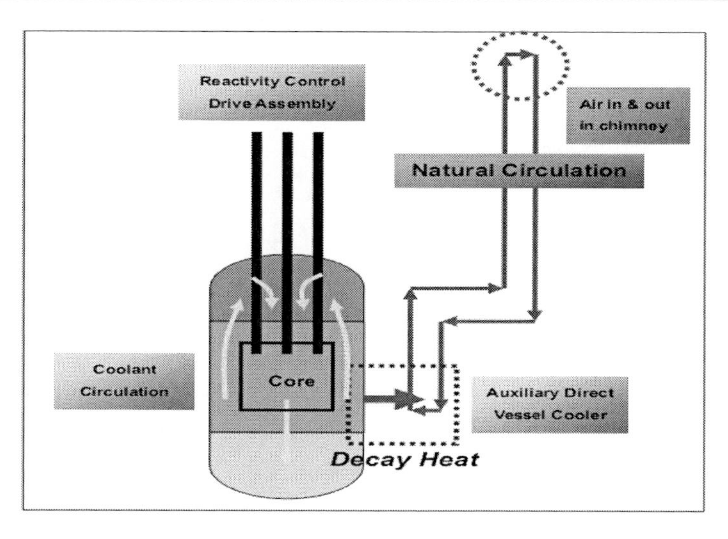

Figure 3. Simplified configuration of natural circulation in VHTR.

Table 2. The flow-chart of the analysis step

Step	Contents
1	Safety margin construction of natural circulation (using mass flow rate)
2	Safety margin construction of Gravity (using Potential energy), Heat conduction (using Temperature), Heat conduction (using Pressure), and Internal stored energy (using Heat transfer rate) by the proportional values to #1
3	Safety margin construction of fuzzy set algorithm
4	Safety margin construction of GAMMA code simulation
5	Probability of failure constructions using safety margins
6	Event/Fault tree construction
7	Data quantification
8	Uncertainty analysis
9	Comparisons

sm(x) = 0 at limit state

sm(x) < 0 for mission failure (1)

Therefore, as Burgazzi postulated (Burgazzi, 2007),

$$Pr_f = Pr(I - A < 0) = \iint_{I-A\leq 0} f_I(I) f_A(A)\, dI dA \tag{2}$$

Using a normal distribution,

$$F(Z) = \frac{1}{\sigma\sqrt{2\pi}} e^{-\frac{(z-\mu)^2}{2\sigma^2}} \tag{3}$$

$$F(Z) \bullet \sigma\sqrt{2\pi} = e^{-\frac{(z-\mu)^2}{2\sigma^2}} \tag{4}$$

From standard normal table and if M is a safety margin, $\mu_M / \sigma_M > 2.33$ $(F(Z) < 10^{-2})$. For nominal value N_n, actual value is N_I, the means of N_A, N_I are μ_A, μ_I, and the variances of N_A, N_I are σ_A^2, σ_I^2. Then,

$$\mu_M / \sigma_M > 2.33 \tag{5}$$

$$(\mu_A - \mu_I) / \left(\sigma_A^2 + \sigma_I^2\right)^{1/2} > 2.33 \tag{6}$$

$$\mu_A > \mu_I + 2.33 \bullet \left(\sigma_A^2 + \sigma_I^2\right)^{1/2} \tag{7}$$

As one can calculate, if $\mu_I = 10\,kg/s$, $\sigma_A = 2$, $\sigma_I = 2$, then, $\mu_A > 16.6kg/s$. In the similar way, other kind of the basic event distribution is constructed using the fuzzy set, which is obtained in Table 3. In case of the triangular form, the distribution of failure frequency can be obtained. For nominal value N_n, actual value is N_I and the means of N_A, N_I are μ_A, μ_I. Then,

$$(\mu_A - \mu_I) / \left|\left(\frac{1}{A} - \frac{1}{B}\right)\right| > 1.0 \tag{8}$$

$$\mu_A > \mu_I + \left|\left(\frac{1}{A} - \frac{1}{B}\right)\right| \tag{9}$$

If $\mu_I = 10\,kg/s$, A= 2, B =3, then, $\mu_A > 10.2kg/s$. So, the maximum safety margin is shown in the membership number = 1.0 as follows,

$$\begin{aligned} &\mathit{Length\,of\,base\,line\,of\,triangles} \\ &= \frac{2\sqrt{5} + 2\sqrt{10}}{2} = \sqrt{5} + \sqrt{10} \end{aligned} \tag{10}$$

Also, for the fuzzy set of the circular form, nominal value is N_n, the actual value is N_I and the means of N_A, N_I are μ_A, μ_I. Then,

$$(\mu_A - \mu_I)/(A-a) > 1.0 \tag{11}$$

$$\mu_A > \mu_I + (A-a) \tag{12}$$

If $\mu_I = 10\,kg/s$, A= 2, a =3, then, $\mu_A > 9.0kg/s$. So, the maximum safety margin is shown in the membership number = 1.0 as follows,

$$\begin{aligned}
&(2 - radius_1 \bullet \cos\theta_1) + \left(\frac{3\sqrt{3}}{2} - radius_2 \bullet \cos\theta_2\right) \\
&= \left(2 - \frac{3\sqrt{3}}{2}\right) + (radius_1 \bullet \cos\theta_1 - radius_2 \bullet \cos\theta_2) \\
&= \left(2 - \frac{3\sqrt{3}}{2}\right) + (2 \bullet \cos\theta_1 - 3 \bullet \cos\theta_2) \\
&= \left(2 - \frac{3\sqrt{3}}{2}\right) + \left(2 \bullet \cos\theta_1 - 3 \bullet \cos\left(\frac{2}{3}\theta_1 + 30^\circ\right)\right)
\end{aligned} \tag{13}$$

Inserting the value $\theta_1 = 90^\circ$, the maximum safety margin is,

$$\left(2 - \frac{3\sqrt{3}}{2}\right) + \left(2 \bullet \cos 90^\circ - 3 \bullet \cos\left(\frac{2}{3}90^\circ + 30^\circ\right)\right) = \left(2 - \frac{3\sqrt{3}}{2}\right) \tag{14}$$

There are four cases for the analysis of the basic event construction for the passive systems which is incorporated with natural circulation. In the other physical variables, the failure frequencies are found out like the mass flow rate in Figure 2. That is to say, the probability of the failure is made using the mean value and the variance. The temperature and pressure can be used for the passive system. The probabilities of the failure for other variables are changed linearly like the mass flow rate. Using the relations of the fuzzy set distributions, the distributions between membership function and safety margin are obtained. As one can see in the above triangle and circular fuzzy sets, the safety margin is related to the slope of the line in the triangle form and to the radius of the circle in the circular form. So, the slope and the radius in the fuzzy set distribution are applied to the standard deviations in the normal distribution. In addition, for the mean value, the position of the sideline in the fuzzy set distribution is related to the position of the center line in the normal distribution. For finding the probability of the failure using fuzzy set distribution, the safety margin in normal distribution is compared with the safety margin in the fuzzy set distributions. Then, the probability of the failure is obtained proportionally by the membership number. One can find the probability of the failure by 1 – 'membership number'. Namely, the maximum value of the probability of the failure has the membership number of 1.0. This is expressed in the Figure 4. The membership number is changed from 0.0 to 1.0. The safety margin value is the relativistic quantity without any unit. The Table 4 shows the relationship between the probability of the filature and the safety margin, which is the case of the Long-term Conduction. For example, the probability of the filature is 0.7 and the safety margin is 6.6 (= $\mu_I - \mu_A$) in the normal distribution.

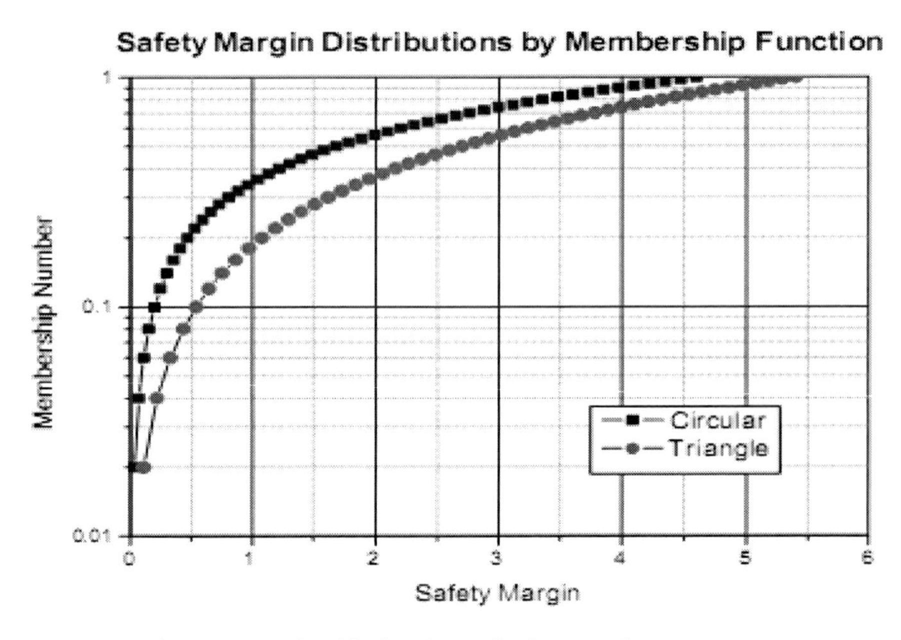

Figure 4. Comparison between membership function and safety margin.

4. Result and Discussion

The probabilities of the failure of the basic events are made in case of the mass flow rate of the natural circulation using the impact-affordability algorithm in the ATWS of VHTR. For finding out the priority of the reliability between the probabilistic method and the fuzzy method, a new value is introduced, which is used for the probability of failure of the event using the GAMMA (GAs Multi-component Mixture Analysis) code. This code was made by the KAERI for the accident analysis of the VHTR (No et al., 2007). The temperature of the fuel is the most important factor to maintain the stability in the VHTR. Therefore, the relationship between the fuel temperature and the failure fraction of the fuel is used. Namely, the failure fraction of the fuel is assumed as the value of the probability of failure. This is made by the relationship of the Figure 5 and 6. The highest temperature of fuel is obtained in Figure 5. The arrow lines show the temperature. This is considered as a safety margin in the mass flow rate of 10 kg/s. This temperature is used to find the failure fraction in Figure6 which is assumed as the probability of the failure (KAERI, 2003). The arrow lines show the failure fraction.

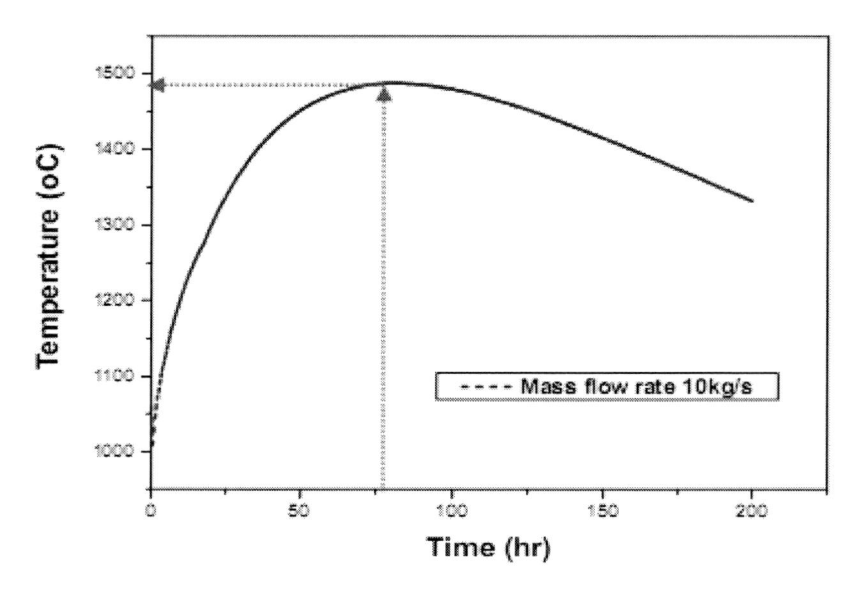

Figure 5. Temperature vs. mass flow rates for maximum fuel temperature using GAMMA code.

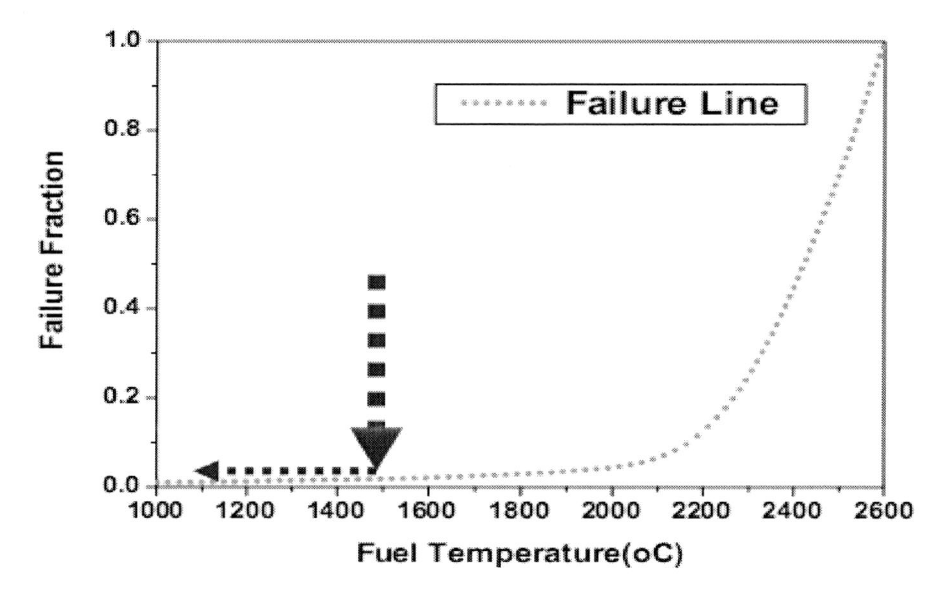

Figure 6. Fuel temperature vs. failure fraction.

For the purpose of the validation of the GAMMA code, it is simulated that the decay heat is removed from the reactor core in a hypothetical accident like the failure of all heat sinks after a reactor trip. The GAMMA code has been validated with the SANA-1 decay heat self-removal test, which is one of the International Atomic Energy Agency (IAEA) Benchmark problems, to investigate the thermofluid characteristics under localized natural convection in a pebble bed. The SANA-1 test apparatus consists of a cylindrical pebble bed having a diameter 1.5 m and a height of 1 m, a central heating element, and a bottom and a top insulator using the various conditions. Additionally, the steady power tests and the power ramp-down and set-up tests are performed. It is possible to examine the several design basis accidents (DBAs) of HTGR by GAMMA (IAEA, 2000).

Table 3. Safety margin by statistical distributions

Normal distribution	Fuzzy set-Triangular	Fuzzy set-Circle
$F_1(z) = \frac{1}{\sigma_1\sqrt{2\pi}} e^{-\frac{(z-\mu_1)^2}{2\sigma_1^2}}$ $\left(-2\ln\left[F_1(z)\sqrt{2\pi}\sigma_1\right]\sigma_1^2\right)^{1/2} = z-\mu_1 - (1)$ $F_2(z) = \frac{1}{\sigma_2\sqrt{2\pi}} e^{-\frac{(z-\mu_2)^2}{2\sigma_2^2}}$ $\left(-2\ln\left[F_2(z)\sqrt{2\pi}\sigma_2\right]\sigma_2^2\right)^{1/2} = z-\mu_2 - (2)$ $(1)-(2)$ $=\mu_2-\mu_1$ $=\left(-2\ln\left[F_1(z)\sqrt{2\pi}\sigma_1\right]\sigma_1^2\right)^{1/2} - \left(-2\ln\left[F_2(z)\sqrt{2\pi}\sigma_2\right]\sigma_2^2\right)^{1/2}$	$F_1(z) = -A\lvert(z-\mu_1)\rvert+1$ $\left\lvert\frac{1}{A}[1-F_1(z)]\right\rvert = (z-\mu_1)-(1)$ $F_2(z) = -B\lvert(z-\mu_2)\rvert+1$ $\left\lvert\frac{1}{B}[1-F_2(z)]\right\rvert = (z-\mu_2)-(2)$ $(1)-(2)$ $=\mu_2-\mu_1$ $=\left\lvert\frac{1}{A}[1-F_1(z)]\right\rvert - \left\lvert\frac{1}{B}[1-F_2(z)]\right\rvert$ $=\left\lvert\left(\frac{1}{A}-\frac{1}{B}\right)-[F_1(z)-F_2(z)]\right\rvert$	$F_1(z) = \sqrt{A^2-(z-\mu_1)^2}$ $\left(A^2-\{F_1(z)\}^2\right)^{1/2} = z-\mu_1 - (1)$ $F_2(z) = \sqrt{a^2-(z-\mu_2)^2}$ $\left(a^2-\{F_2(z)\}^2\right)^{1/2} = z-\mu_2 - (2)$ $(1)-(2)$ $=\mu_2-\mu_1$ $=\left(A^2-\{F_1(z)\}^2\right)^{1/2} - \left(a^2-\{F_2(z)\}^2\right)^{1/2}$

Table 4. Probability of failure vs. safety margin

	Normal Dist.	Fuzzy -Triangle	Fuzzy-Circle	Gamma code
Prob.of failure	0.7	0.9909	0.9545	N/A
Safety margin $(\mu_I - \mu_A)$	6.6	0.2	1	N/A

The Figure 7 is used for the quantification of the event/fault tree (USNRC, 2003). The red color part is related to the passive system in the natural circulation. For the basic event of the active system part, a modified event likelihood of occurrence is used for the ATWS. The fault tree of Figure 7 shows that the Core Power Failure is connected to the initiation event of the SCS Failure to Start and the Flow Coast Down and Power eq.

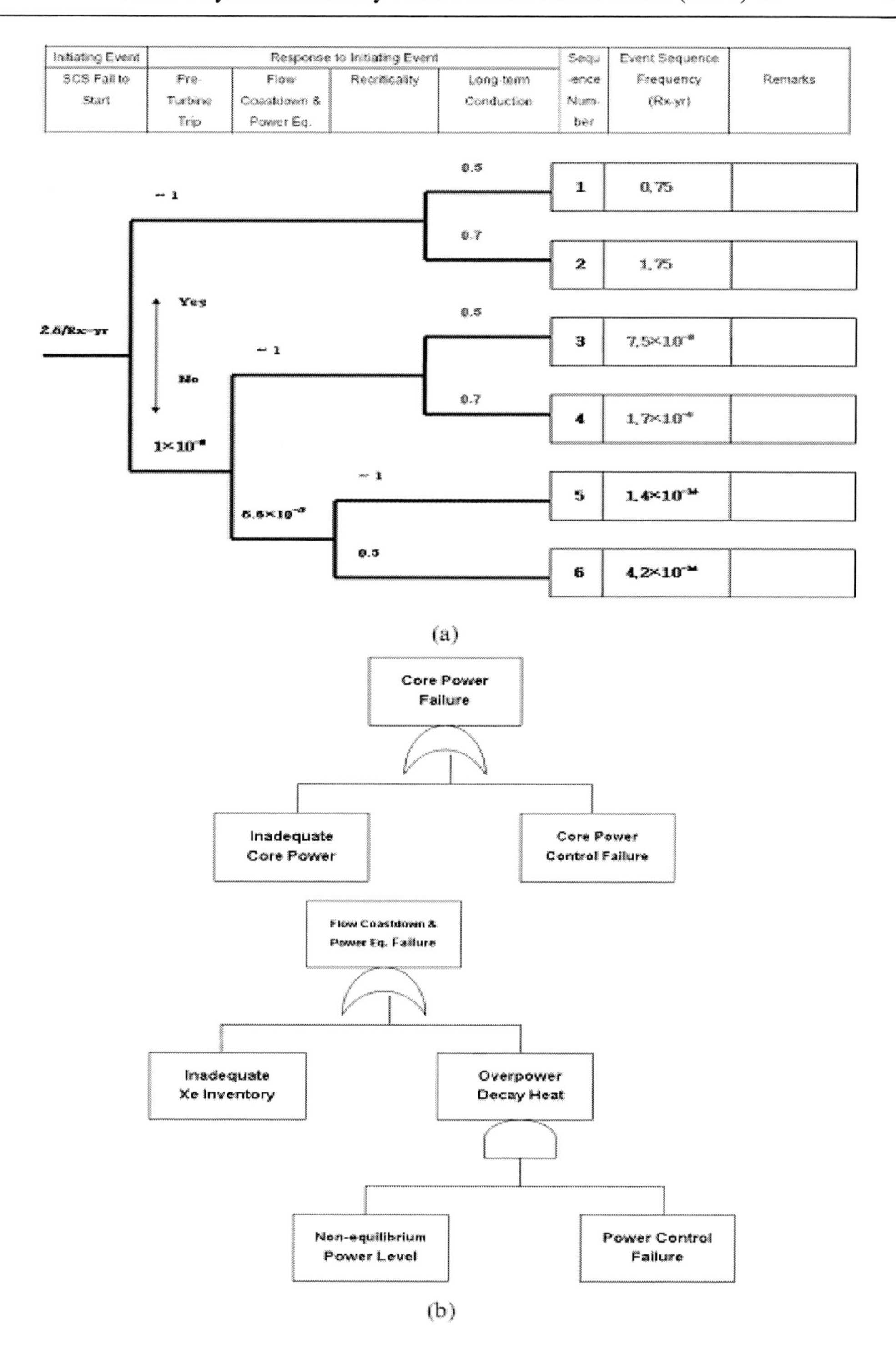

Figure 7. Event/fault tree for VHTR (a) Event tree, (b) Fault tree.

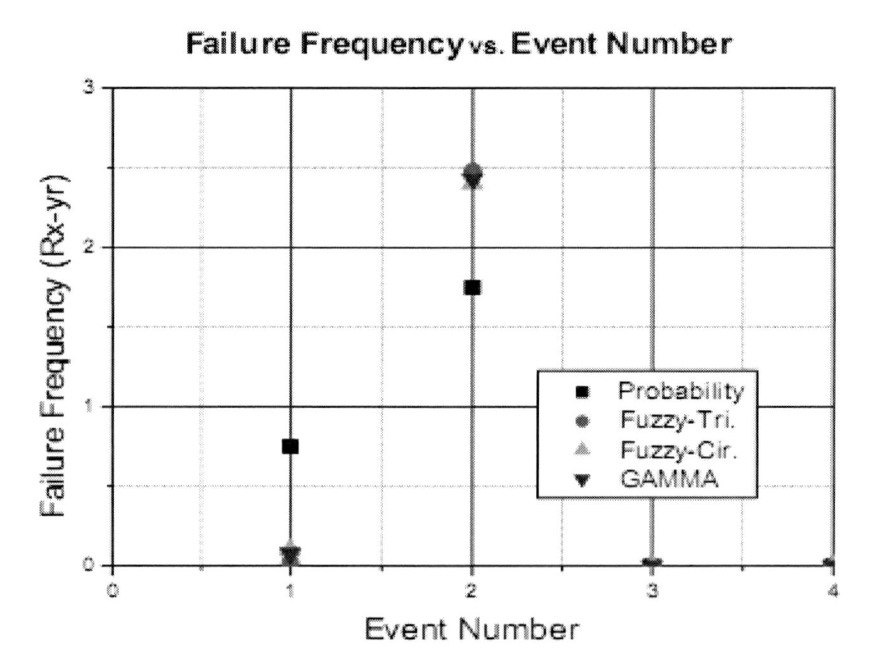

Figure 8. Comparisons of several methods.

Failure is connected to the Flow Coast Down and Power eq. This case is treated as the conventional method which is used in the event/fault tree quantification without using of the impact-affordability algorithm.

The data in Table 5 are used for the non-passive system parts as the probability of the failure, which is modified from the event likelihood of occurrence based on SECY-93-092 (USNRC, 1993). Table 6 shows the quantifications of the propagation using the impact-affordability algorithm (ATWS case).

Propagations of the fuzzy set distribution have lower values than those of the probability set except case #2. So, the Figure 8 shows that the failure frequencies of the fuzzy set are conservative than those of the probability set, where the probabilistic and fuzzy methods are compared with the thermohydraulic simulation of the GAMMA code. The failure frequencies of case #1 and #2 show that the values of the fuzzy set are similar to those of the GAMMA code. This means that values of the fuzzy set have much more realistic comparing to those of the probability set. Otherwise, there is no significant difference in case #3 and #4. The values of the GAMMA code simulation are used as the physical calculation which is compared with the statistical values of the probability and fuzzy sets. The uncertainty analysis is done Table 7, 8, 9 and 10.

Conclusions

The dynamical safety assessment has been developed newly for a reactor. This study shows the non-linear algorithm could substitute with the probabilistic descriptions. The restrictions of the conventional probabilistic analysis like the mean value and the standard deviation are changed to the some geometrical configuration as the slope and the radius in the membership function. This makes the conceptions of the physical variables be expressed much more easily and exactly manipulated by the operator. Therefore, the priority of the non-linear algorithm can be found. The particular meanings of this study are as follows,

- New kind of PSA for the passive system is investigated using the impact-affordability algorithm of the event.
- A safety margin is obtained successfully using statistical variables (mass flow rate and other variables).
- The restrictions of the probabilistic distributions are modified to the simpler ways as the geometrical expressions.
- The test for the conventional statistical distribution is compared with the non-linear complex distribution.
- A new PSA algorithm can be used in the regulation standard.
- The impact-affordability algorithm can be used to active as well as passive systems.

Using the interpretation of the fuzzy set distributions, the quantity of the physical variables can be showed with the linguistic expression of the operator. Since the fuzzy set theory is originated from the linguistic expression of the operator, the human error could be reduced in some active systems. Additionally, other kind of the complex algorithm like the neural network or chaos theory could be used for the data quantification in PSA.

Table 5. Modified event likelihood of occurrence based on SECY-93-092

Event	Frequency of Occurrence
Likely events	$0 \sim 10^{-2}$ / plant-year
Non-likely events	$10^{-2} \sim 10^{-4}$ / plant-year
Extremely non-likely events	$10^{-4} \sim 10^{-6}$ / plant-year
Very rare events	$< 10^{-6}$ / plant-year

Table 6. Failure frequency of event for ATWS (Rx-yr) (Natural circulation related events)

Event	Normal Dist.	Fuzzy -Triangle	Fuzzy-Circle	GAMMA code
1	7.50×10^{-1}	2.00×10^{-2}	1.00×10^{-1}	7.50×10^{-2}
2	1.75×10^{0}	2.48×10^{0}	2.40×10^{0}	2.43×10^{0}
3	7.50×10^{-6}	2.30×10^{-8}	1.10×10^{-7}	7.50×10^{-8}
4	1.70×10^{-5}	2.50×10^{-6}	2.40×10^{-6}	2.43×10^{-6}

Table 7. Uncertainty for ATWS (Rx-yr) in Normal. Dist

Event	Mean	Median	25th %	75th %
1	7.50×10^{-1}	1.05×10^{0}	2.09×10^{-2}	1.53×10^{0}
2	1.75×10^{0}	2.16×10^{0}	1.28×10^{0}	3.03×10^{0}
3	7.50×10^{-6}	7.12×10^{-5}	3.57×10^{-5}	1.07×10^{-4}
4	1.70×10^{-5}	2.91×10^{-5}	1.45×10^{-5}	4.36×10^{-5}

Table 8. Uncertainty for ATWS (Rx-yr) in Fuzzy-Triangle

Event	Mean	Median	25 th %	75 th %
1	2.00×10^{-2}	2.73×10^{-2}	1.65×10^{-2}	3.81×10^{-2}
2	2.48×10^{0}	2.86×10^{0}	1.82×10^{0}	3.89×10^{0}
3	2.30×10^{-8}	1.90×10^{-7}	9.53×10^{-8}	2.85×10^{-7}
4	2.50×10-6	3.60×10-5	1.80×10-5	5.38×10-5

Table 9. Uncertainty for ATWS (Rx-yr) in Fuzzy-Circle

Event	Mean	Median	25 th %	75 th %
1	1.00×10^{-1}	1.31×10^{-1}	7.39×10^{-2}	1.88×10^{-1}
2	2.40×10^{0}	2.77×10^{0}	1.77×10^{0}	3.76×10^{0}
3	1.10×10^{-7}	9.49×10^{-7}	4.76×10^{-7}	1.42×10^{-6}
4	2.40×10^{-6}	3.48×10^{-5}	1.74×10^{-5}	5.22×10^{-5}

Table 10. Uncertainty for ATWS (Rx-yr) in GAMMAcode

Event	Mean	Median	25 th %	75 th %
1	7.50×10^{-2}	8.78×10^{-2}	5.51×10^{-2}	1.20×10^{-1}
2	2.43×10^{0}	2.81×10^{0}	1.78×10^{0}	3.83×10^{0}
3	7.50×10^{-8}	6.32×10^{-7}	3.17×10^{-7}	9.46×10^{-7}
4	2.43×10^{-6}	3.47×10^{-5}	1.73×10^{-5}	5.20×10^{-5}

Acknowledgments

Author thanks to the Drs. H. S. Lim and S. J. Han in Korea Atomic Energy Research Institute (KAERI) in Korea for their research discussions.

References

Bianchi, F., Burgazzi, L., D'auria, F., Galassi, G. M., Ricotti, M. E., ORIANI, L., 2001.

Burgazzi, L., 2007. Thermal-hydraulic passive system reliability-based design approach. *Reliability Engineering and System Safety* 92, 1250-1257.

Evaluation of the reliability of a passive system. *International conference of nuclear energy in central Europe* 2001.

IAEA, 2000. Heat Transfer and Afterheat Removal for Gas Cooled Reactors under Accident Conditions. *IAEA-TECHDOC*-1163.

KAERI, 2003. *Basic Study on High Temperature Gas Cooled Reactor Technology for Hydrogen Production*. KAERI/RR-2288/2002, Korea.

Korea Atomic Energy Research Institute, 2007. *Experiment Database:I-NERI Final Project Technical Report*. Korea.

Nayak, A. K., Sinha, R. K., 2007. Role of passive system in advanced reactor, *Progress in Nuclear Energy* 49, 486-498.

No, H., et al., 2007. Multi-component diffusion analysis assessment GAMMA code and improved RELAP5 code. *Nucl. Eng. and Design*. 237, 997-1008.

USNRC, 1993. SECY-93-092, PRA attachment 4. *USNRC Commission Papers*. USA.

USNRC, 2003. *Regulatory Effectiveness of the Anticipated Transient Without Scram Rule*, NUREG-1780. USA.

Zio, E., Baraldi, P., 2003. Sensitivity analysis and fuzzy modeling for passive systems reliability assessment. *Annals of Nuclear Energy* 31, 277-301.

In: Artificial Intelligence
Editor: Brent M. Gordon, pp. 143-150
ISBN 978-1-61324-019-9

Chapter 7

EMERGENT TOOLS IN AI

Angel Garrido
Facultad de Ciencias (UNED), Madrid, Spain

ABSTRACT

The historical origin of the Artificial Intelligence (AI) is usually established in the Darmouth Conference, of 1956. But we can find many more arcane origins [1]. Also, we can consider, in more recent times, very great thinkers, as Janos Neumann (then, John von Neumann, arrived in USA), Norbert Wiener, Alan Mathison Turing, or Lofti Zadeh, for instance [12, 14]. Frequently AI requires Logic. But its Classical version shows too many insufficiencies. So, it was necessary to introduce more sophisticated tools, as Fuzzy Logic, Modal Logic, Non-Monotonic Logic and so on [1, 2]. And we need a new Mathematics, more adapted to the real world, which may comprises new areas, as Fuzzy Sets, Rough Sets, and Hybrid Systems; for instance, analyzing Knowledge with Uncertainty. Among the things that AI needs to represent are *Categories*, *Objects*, *Properties*, *Relations between objects*, *Situations*, *States*, *Time*, *Events*, *Causes and effects*, *Knowledge about knowledge*, *and so on.* The problems in AI can be classified in two general types [3, 5], *Search Problems* and *Representation Problems.* On this last "peak", there exist different ways to reach their summit. So, we have [4] *Logics, Rules, Frames, Associative Nets, Scripts*, and so on, many times connected among them. We attempt, in this paper, a panoramic vision of the scope of application of such emergent methods in AI.

Keywords: Knowledge Representation, Heuristics, Rough and Fuzzy Sets, Bayesian Networks, AI in Medicine.

1. REPRESENTATION PROBLEMS

We can use a series of resources [4, 5] to approach the problems in A I, *Logic, Rules, Associative Nets, Frames* and *Scripts.* The election between these methods depends on the particular characteristics of the problem and also on our expectation about the type of solution

[3]. In many cases, we take two o more tools at a time, as in the case of the Frame System, with the participation of Rules, and so on.

2. Rules

Concerning the usual way of appearance of Rules [3, 4, 5], as *RBS* (acronym of *Rule Based Systems*), we need four elements: *Interface of Usuary (IU)*, which will be very useful for the interaction with the usuary; *Motor of Inference (MI)*, responsible for the control of the flow of information between the modules; *Base of Facts* or *Base of Affirmations (BF or BA)*, that contains the initially known facts and those created during the process; and *Base of Knowledge (BK),* which contains the Rules used for the Representation of knowledge, into a determined Domain. There exists a two-way flow: *from the MI to IU,* and *from MI to BA,* but only one between *BK* and *MI,* and never in the reverse sense, except if we accept the system´s capacity of learning.

3. Inference in SBR

Such *Inference* consists in establishing the certainty of some statement, from the disposable information into Base of Facts and Base of Knowledge. We can use two methods, *concatenation going forward*, or *concatenation going backwards.* In the first case, we depart from Rules with verified affirmations in their antecedent, advancing through the affirmations which we find in their consequents. Whereas in the second case, we depart of Rules verified in certain consequents (all the consequents must be also verified in this sense), and we turn back to the antecedent. This converts its affirmations in new sub-objectives for the proof, searching for Rules where appear in their consequent, and so on.

The Rules shows a great advantage compared to the Classical Logic [5]. In Classical Logic, as you known, the Reasoning was Monotonic, with inferences without contradiction with the pre-existing, in RBS. Nevertheless, in the RBS, we may delete facts or affirmations from the Base of Facts, according to the new inferences. This makes the Reasoning Non-Monotonic, because we can modify the conclusion. Then, a question arises: what should we do with the conclusion of the affirmation now invalided? For this problem [3], we need to introduce the concept of *Type of Dependence of a Rule*, which can be *Reversible:* if we delete the affirmations, then we automatically delete the above inferred facts, or *Irreversible:* the facts, once inferred, are not deleted or changed.

And in the case of some applicable rules to time, which of them should be first executed? Such Rules constitute, for each step, the *Conflict Set* (obviously, a dynamic set). The subjacent decision problem is called *Resolution of Conflicts* or *Control of Reasoning.* There exist different strategies, to elect a Conflict Set, such as *Ordering of Rules*; *Control of Agendas*; *Criterion of Actuality*, and *Criterion of Specificity.* About the first and the second, the commentaries are unnecessary: they consist in the disposition of the Rules in the order as must be executed. The *Criterion of Actuality* consists in applying first the Rules in whose Antecedent there exists the up-to-date information. The Motor of Inference must be charged of the control of their respective moments. The *Criterion of Specificity* leads to executing the more specific Rules first, that is, those with more facts in their antecedent. So, between R_1: *if a, then b*, and R_2: *if a and d, then c*, we must select R_2, because it is more specific than R_1.

We also have need the *Mechanisms of Control in RBS.* So, by using*Mechanism of Refractority:* we are prevented from executing again a Rule already utilized, unless there is no other information such (in general, anomalous) case; *Rule Sets*: they allows the activation or neutralizing of Block´s Rules; *Meta-Rules*: they are rules which treat (or reason about) other Rules. Such Meta-Rules can collaborate in the Control of Reasoning, with the change or assignation of priorities to different Rules, according to the evolution of the circumstances.

4. Frames

They constitute the most general and most integrating method among all the Representation Procedures [3, 5]. They us to introduce some different elements. For instance, Rules in Frame Systems. Such System are usually denoted as *FS.* We must distinguish between *Facets,* as properties of the Field, and *Devils,* as procedures associated to the Frame System.

5. Scripts

They are structures of knowledge [1, 3, 4, 5] which must organize the information relative to dynamical stereotyped situations, that is, always, or almost always identical sequence of steps, or at least very similar. For instance, going to a certain cinema or a certain big store. The words and the subjacent ideas remind one of movies.

The *elements of a Script* can be *Scenes, Roles, Objects, Places, Names, Conditions, Instruments,* and *Results.* Its signification is obvious from the name: for instance, the Scenes must be events described sequentially, each scene being necessary for the realization of the subsequent one. With Results, we mention the facts we have obtained, once we have finished the sequence described in the Script.

6. Searching Methods

We will distinguish between Blind Search Procedures and Heuristic Procedures. In the first case, the oldest of them, it is possible to apply *Breadth Search* and *Depth Search.* But the problems associated to Combinatorial Explosion occur when the ramification index, or branching factor (the average cardinal of the successors of each node) increase beyond reasonable bounds.

For this reason, a more efficient search procedure is required, such as the introduction of *heuristic functions,* which give the estimation of the distance among the actual node and the final node. Thus, we chose to cal it *Heuristic Search.*

7. INTRODUCTION TO FUZZINESS

We define the "world" [4] as a complete and coherent description of how things are or how they could have been. Often, in problems related to the "real world", which is only one of the "possible worlds", the Monotonic Logic does not work often. But such a type of Logic is classically used in formal worlds, such as Mathematics. It is a real problem, because the "common sense" logic is non-monotonic, and this is our usual logic. We can see the more essential foundations of Fuzzy Theory in books as [2], [3] or [13].

An element of the Universe, U, can belong more or less to an arbitrary set C. It can belong to C in different degrees. From 0, when it does not belong at all to C, to 1, when it belongs totally to C. Or in any intermediate degree, like: 0.5, 0.3, 0.1..., but always between 0 and 1, both values included in their range. Such "membership degree" value can be assigned by an adequate "membership function", whose range is the closed unit interval, [0, 1]. So, the application can be expressed as in [10], by $f: C \rightarrow [0, 1]$. But information is given about the "membership degree", of such element, x, of the universe U, to the set C. In a Classical Set, therefore, the range of f should be reduced to the set {0, 1}. Given n universes of the discourse, we define a fuzzy relation, R, as a membership function that associates each n-tuple, a value of the unit closed interval, [0, 1]. The fuzzy relation, R, can be defined through such "membership function". In this way, we have 0R, ..., (1/3) R, ..., (1/2)R, ..., 1R.

The Cartesian product of two fuzzy sets, F and G, will be a fuzzy binary relation, through the minimum between the membership degrees. Sometimes, it is very useful to symbolize each fuzzy relation as one matrix, where the entries can be any real number between 0 (not related) and 1 (totally related, or simply, related). There exists a clear analogy between the composition of fuzzy relations and the product of matrices. To show this connection, it is sufficient to establish the correspondence: one +, one max; and one • one min. For this reason, the composition of fuzzy relations can also be called "max-min matrix product". As a particular case of the previous definition for the composition between fuzzy relations, we can introduce the composition between a fuzzy set and a fuzzy relation.

This can be very useful in the "Fuzzy Inference" [7], where we attempt to obtain new knowledge from the only already available. Obviously, in such a case, the fuzzy set can be represented by one row matrix, or a column matrix, depending on the order in the product.

The usual properties of the classical relations can be translated into fuzzy relations, but the transitive will be modified. R is *Reflexive:* R(x, x) = 1, for each x in the set C, into the universe. According to this, each element would be totally related with itself, when R is reflexive. R is *Symmetric:* If R(x, y) = R(y, x), for each pair (x, y). Therefore, the principal diagonal acts as a mirror, in the associated matrix. R is *Transitive:* Not in the usual way for relations or associated matrices, but now it is that $R(x, z) \geq \max(\min\{R(x, y), R(y, z)\})$ occurs.

All these mathematical methods can be very useful in Fuzzy Logic and in many branches of Artificial Intelligence. We can introduce new generalized versions of Classical Logic. So, *Modus Ponens Generalized, Modus Tollens Generalized* or *Hypothetic Syllogism.*

To each Fuzzy Predicate, we will associate a Fuzzy Set, the defined by such property, that is, composed by the elements of the Universe such that totally or partially verify such condition. For example, we can prove that the class of fuzzy sets with the operations $\cup$, $\cap$ and c (path to the complementary) does not constitute a Boolean Algebra, because neither the

Contradiction Law nor the Middle Excluded Principle work in it. Geometrical and algebraic proof is easy, by a counterexample: it suffices to take an element with membership degree that belongs to the open unit interval.

8. ROUGHNESS

The concept of *Rough Set* was introduced by the Polish mathematician Zdzislaw Pawlak in 1982. Some theoretical advances with the corresponding applications have been emerging since then [2]. It is possible to apply Rough concepts to astonishing purposes, as will be the prediction of financial risk, but also in voice recognition, image processing, medical data analysis and so on.

Taking object, attribute or decision values, we will create rules for them: upper and lower approximations and boundary approximation. Each object is classified in one of these regions. For each rough set, $A \subseteq U$, we dispose of *Lower Approximation of A*, as the collection of objects which can be classified with full certainty as members of A, and *Upper Approximation of A*, as the family of objects which may possibly be classified as members of A. Obviously, this class is wider than the aforementioned, containing between both the Rough set. *Rough Set Theory is a model of Approximate Reasoning.* According to this, we will interpret knowledge as a way to classify objects. We dispose of U, the universe of discourse, made up of objects, and an equivalence relation on U, denoted R. The procedure is to search for a collection of subsets in U (categories), so that all the elements of the same class possess the same attributes with the same values. So, we obtain a covering of U by a set of categories.

The elementary knowledge is encoded in a pair *(U, R),* made up of *"elementary granules of knowledge".* They constitute a partition in equivalence classes, into the quotient set, U/R. Given two elements, it is possible to determine when they are mutually indiscernible. In this case, we call this the *Indiscernibility Relation.* Therefore, it is possible to introduce the application which assigns to each object its corresponding class. Then, such indistinguishability allows us to introduce the *Fibre of* a_R, defined by the aforementioned relation R. So, the collection of such fibres, in the finite case, produces a union: this union of fibres is called a *granule of knowledge*. The pair (U, R) will be a *Knowledge Base*.

We say that an object, or category, is R-rough, if it is not R-exact. For each R-rough set, $Y \subseteq U$, we define two associate R-exact sets, the *R-lower approximation of Y,* and the *R-upper approximation of Y.* So, we can represent the Rough set, Y, through the pair ($\underline{R}$ Y, $\overline{R}$ Y). Observe that $\underline{R}\, Y \subseteq Y \subseteq \overline{R}\, Y$, and furthermore, Y is R-exact $\Leftrightarrow \underline{R}\, Y = \overline{R}\, Y$.

Given a Knowledge Base, K ≡ (U, R), we will take the collection of classes E_K = *{R - exact sets on U}*, which is closed with respect to the usual set operations $\cup$, $\cap$ and c. It verifies the known properties of a Boolean Algebra. More concretely, we can say a *Field of Sets.* But it is not the case when we deal with R-rough sets. Because, for instance, the union of two R-rough sets can be a R-exact set. The coincidence of this Rough Set Theory with the Classical Theory of Sets occurs when we only work with R-exact sets.

An interesting generalization of Rough Set will be the *Generalized Approximation Space,* denoted *GAS.* It consists of a triple: (U, I, ν), where *U* will be the *Universe; I,* the *uncertainty function, I:* $U \rightarrow P(U)$, and ν the *Rough Inclusion Function.* An example of this type of Rough Inclusion Function will be the *Generalized Rough Membership Function.* So, given any subset, we have both GAS - approximations, *lower-approximation* and *upper-approximation.*

9. Comparison between Fuzziness and Roughness

These names us mislead into believing that they are referring to the same concept. But they are very different approaches to uncertainty in the set of data. It depends of the nature of vagueness in the problem, or the convenience in applications. Both resources cover distinct aspects of the Approximate Reasoning. For this reason, both paradigms address to solve the Boundary Problem in Non-Crisp cases. Dubois and Prade [2] establish the mutual relationship between *Rough Fuzzy Set* and *Fuzzy Rough Set.* In the first case we will pass from fuzzy sets, through filtering, by the classical equivalence relations to quotient spaces, which are fuzzy sets. Whereas, in the second case we imitate the rough set approximation, but now with fuzzy similarity (instead of equivalence) relations.

We work into the collection of fuzzy sets on U, endowed with the operations: max and min. So, {Fuz (U, [0, 1]), max, min}. This produces a Zadeh Lattice. And it provides the path to complementary operator, {Fuz (U, [0, 1]), max, min, c}. It will be a Brouwer-Zadeh Lattice.

This lets us introduce the Rough Approximation to Fuzzy Sets. Our actual purpose is double: given $A \in Z$ (U), we can induce a fuzzy set in U / R, by A, and reach the approximation of A, relative to R, according to the Rough Set Theory.

The notion of Fuzzy Rough Set is dual to the above concept. We consider newly the family of fuzzy sets in the universe U, with values in the closed unit interval, Fuz (U, [0, 1]). We need to analyze the fuzzy notion of equivalence relation and then, the fuzzy partition induced. Regarding the equivalence relation, the closest concept is the T-Fuzzy Similarity Relation.

In the past, the relationship between Fuzzy and Rough concepts were studied by some mathematicians and computer scientists, as Pawlak, Nakamura, Pedrycz [7], Dubois and Prade, Pal and Skowron, and many others.

10. Networks

As far as Nets are concerned, the more actual studies to deal with Bayesian Nets, also called Belief Networks [3, 5]. Before their apparition, the purpose was to obtain useful systems for the medical diagnosis, by classical statistical techniques, such as the Bayes´s Rule or Theorem.

A *Bayesian Net* is a pair (G, D), with G a directed, acyclic and connected graph, and D a distribution of probability (associated with the participant variables). Such distribution, D, must verify the *Property of Directional Separation,* according which the probability of a variable does not depend upon their not descendant nodes.

The *Inference in BNs* consists in establishing on the Net, for the known variables, their values and or the unknown variables, their respective probabilities. The objective of a

Bayesian Network, in Medicine, is to find the probability of success with which we can give an exact diagnosis, based on known symptoms. We need to work with the following *Hypothesis: Exclusivity, Exhaustivity, and Conditional Independence.* According to the *Hypothesis of Exclusivity,* two different diagnoses cannot be right at the same time. With the *Hypothesis of Exhaustivity,* we suppose at our disposition all the possible diagnosis. And by the *Conditional Independence,* the thing found must be mutually independent to a certain diagnosis.

The initial problem with such hypothesis was the usual: their inadequacy to the real world. For this, we need to introduce the Bayesian Networks (BNs). In certain cases, as in the vascular problem of the predisposition to heart attack, there already exist already reasonable Systems of Prediction and Diagnosis, such as the DIAVAL Net. From these procedures springs a new and useful sub-discipline called *Medical artificial intelligence* (MAI, in acronym).

There are many different types of clinical tasks to which Expert Systems can be applied, as Generating alerts and reminders; Diagnostic assistance ; Therapy planning; Agents for information retrieval; Image recognition and interpretation, and so on. In the fields of treatment and diagnosis, A I possesses very important realizations, giving us for instance the following tools: PIP (1971), at MIT; MYCIN (1976), a Rule-Based System, due to Stanford University, works on infectious diseases; CASNET (1979) is due to Rutgers University and works on ophtalmological problems; INTERNIST (1980), due to Pittsburgh, on inner medicine; AI/RHEUM (1983), at Missoury University, on Reumathology; SPE (also 1983), at Rutgers, analyses the electrophoresis of proteins; TIA (1984), at Maryland, on the therapy of ischemic attacks, and many others.

REFERENCES

[1] Barr, A., and Feigenbaum, E. A. (1981). *The Handbook of Artificial Intelligence,* Volumes 1-3. William Kaufmann Inc.

[2] Dubois, D., and Prade, H. [eds.] (2000), *Fundamentals of Fuzzy Sets*, Kluwer Publ., Dordrecht, Netherlands.

[3] Fernández Galán, S., et al. (1998). *Problemas resueltos de Inteligencia Artificial Aplicada. Búsqueda y Representación.* Addison-Wesley. Madrid.

[4] Garrido, A. (2004). *Logical Foundations of Artificial Intelligence.* "FotFS V", Institut fur Angewändte Mathematik, Friedrich-Wilhelm Universität, University of Bonn.

[5] Mira, J., et al. (1995). *Aspectos Básicos de la Inteligencia Artificial.* Editorial Sanz y Torres, Madrid.

[6] Mordeson, J. N.; Malik, D. S., and Cheng, S. C. (2000). *Fuzzy Mathematics in Medicine*. Physica Verlag, Heidelberg.

[7] Pedryck, W. (1993). *Fuzzy Control and Fuzzy Systems*, Wiley and Sons Co., New York, 1993.

[8] Pólya, G. (1987). *Combinatorial enumeration of groups, graphs...* Springer Verlag.

[9] Searle, J. (1980). *Minds, Brains and Programs.* The Behavioral and Brain Sciences, 3, pp. 417-424.

[10] Szczepaniak, P. S.; Lisoba, P. J. G., and Kacprzyk (2000). *Fuzzy Systems in Medicine.* Physica Verlag, Heidelberg.

[11] Torres, A. and Nieto, J. J. (2006). *Fuzzy Logic in Medicine and Bioinformatics.* J. of Biomedicine and Biotechnology, Vol. 2006, pp. 1-7, Hindawi Publishers.

[12] Turing, A. (1950). *Computing machinery and intelligence.* Mind 59, pp. 433-60.

[13] Wang, Z. (1997). *A Course in Fuzzy Systems and Control*, Prentice-Hall, New Jersey.

[14] Wiener, N. (1947). *Cybernetics.* MIT Press and John Wiley. New York.

In: Artificial Intelligence
Editor: Brent M. Gordon, pp. 151-160
ISBN 978-1-61324-019-9

Chapter 8

NEURAL NETWORKS APPLIED TO MICRO-COMPUTED TOMOGRAPHY

Anderson Alvarenga de Moura Meneses, Regina Cely Barroso and Carlos Eduardo deAlmeida
Rio de Janeiro State University, Brazil

ABSTRACT

In this chapter, we review advances of Artificial Intelligence (AI) applications, namely Artificial Neural Networks (ANNs), for the micro-Computed Tomography, a biomedical imaging technique. AI, and particularly ANNs, have yielded outstanding results in biomedical imaging, especially in image processing, feature extraction, classification and image interpretation. ANNs become a great ally for biomedical image analyses, in cases which traditional imaging approaches are not sufficient to detect specific characteristics, intrinsic patterns, or when computer-aided diagnosis must be sensitive to details associated to certain level of perception in the visualization. We discuss concepts related to this application of ANNs (training strategies), presenting results of the successful use of this technique.

Keywords: Artificial Intelligence, Artificial Neural Networks, Biomedical Imaging, Image Processing and Analysis.

1. INTRODUCTION

Researchers in biomedical sciences readily realized that the application of Artificial Intelligence (AI) might bring enormous benefits in both diagnosis and treatment of patients [1]. Besides the investigation on AI techniques concerning the development of rule-based expert systems for computer-aided diagnosis in the 70s and 80s, e.g. [2, 3], the application of AI to medical imaging also became a major field of investigation, focusing on several imaging modalities such as ultra-sound [4] and microscopy [5].

With the evolution and dissemination of the neural computing, the use of Artificial Neural Networks (ANNs) [6] in biomedical imaging increased. ANNs are AI models based upon the relationships among neurons, synapses and learning, and are used for tasks such as regression, classification and pattern recognition. For example, in 1988, Egbert, Rhodes and Goodman [7] applied ANNs to biomedical infra-red imaging for the classification of tumor stages in laboratory mice.

Nevertheless, the current application of ANNs to medical imaging may encompass tasks such as preprocessing, data reduction, segmentation, object recognition, image understanding and optimization, as in the taxonomy adopted by Egmont-Petersen, de Ridder and Handels in [8], where the authors review the increasing number of solutions involving ANNs in image processing and analysis.

Progress in medical imaging such as Computed Tomography (CT) [9] has allowed breakthroughs in biomedical imaging of biological tissue. The CT enables the observance of characteristics and aspects in three dimensions that would not be possible with conventional X-ray techniques. The micro-Computed Tomography (μCT) obtained with Synchrotron Radiation (SR) [10] such as the one used in this research provides an spatial resolution of approximately 14μm due to an extremely coherent and collimated beam. This and other characteristics such as the phase-contrast imaging [11, 12], a physical phenomenon that causes the enhancement of edges of, for example, soft biological tissues in the images, makes SR-μCT very interesting for biomedical purposes, with applications in mammography [13], bone mass assessment [14], and biological morphology [15]. Therefore, advances in imaging modalities demand the investigation of novel algorithms, as well as well-known techniques.

In the present chapter, we review the proposal of training strategies, which are combinations of training algorithms, sub-image kernels and symmetry information. Investigations demonstrated that Feed Forward (FF) ANNs are suitable for the recognition of bone pixels in SR-μCT medical images without any sort of preprocessing such as normalization, filters or binarization [16, 17]. We also present results of a study comprising the statistical analysis and assessment of different architectures, training algorithms and symmetry strategies of FF-ANNs for SR-μCT images.

The remainder of this chapter is organized as follows. A brief discussion about the SR-μCT for biomedical imaging is presented in section 2, with an overview of the instrumentation. The training strategies for ANNs used in the segmentation of SR-μCT biomedical images are discussed in section 3. Section 4 presents the validation methodology. Computational experimental results are presented in Section 5, and the discussion, in section 7. Finally, conclusions are in section 7.

2. Synchrotron Radiation Micro-Computed Tomography for Biomedical Imaging

High resolution imaging techniques such as the SR-μCT play an important role for the progress of diagnosis based on medical imaging. In particular, the SR-μCT has great potential and the study of techniques and application is therefore fundamental for future advances, since, according to Chappard et al. [14], "μCT is only at its beginning".

SR facilities such as the Elettra Laboratory (Trieste, Italy) provide important characteristics for the acquisition of medical images. The image of interest (Figure 1) was obtained at the Synchrotron Radiation for MEdical Physics (SYRMEP) beam line at the Elettra Laboratory, which is designed for *in-vitro* samples X-Ray imaging. The beam is laminar and extremely collimated. The SR beams are, to a good extent, coherent, allowing higher sensitivity and better spatial resolution and this is an important factor for innovative imaging techniques. Moreover, the high brilliance of the synchrotron light allows the use of monochromatic radiation, i.e., the selection of single photon energy. Monochromaticity avoids beam hardening and therefore turns a reconstructed radiography in a quantitative mapping of X-ray attenuations.

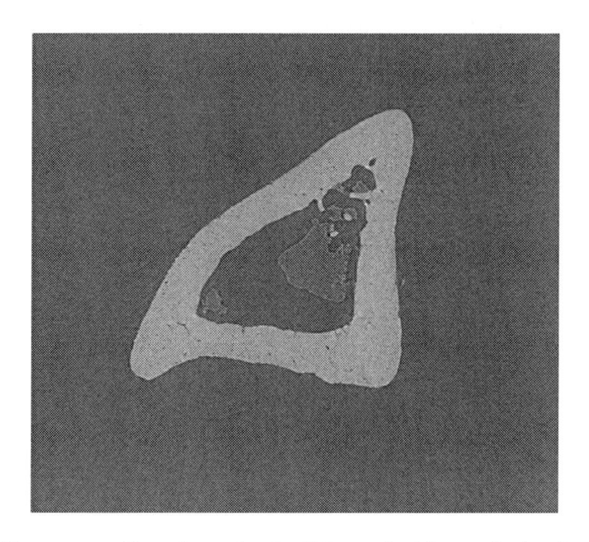

Figure 1. The image of interest: a slice of a rodent's tibia cortical bone obtained with Synchrotron Radiation μCT.

Despite the enhancement of biological tissue imaging due to the monochromaticity and coherence of SR beams, radiological imaging based only on absorption effects results in poor image contrast [13]. Notwithstanding, the information carried by the phase of X-Ray wavefield, or *phase effects*, may be exploited by techniques that convert it into image contrast, enhancing detail visibility [12]. Lewis [11] reviewed the three experimental techniques used in order to obtain images with contrast enhancement based on phase-shift effects: X-Ray interferometry, Diffraction Enhanced Imaging and in-line holography (also called free-space propagation or merely *phase contrast* imaging [12]).

The SYRMEP beam line provides a monochromatic laminar-section X-ray with a maximum area of about 160 x 5 mm^2 at 20 keV, at a distance of about 23 m from the source. The system consists of a Si (111) crystal working at Bragg configuration. The useful energy range is 8-35 keV. The intrinsic energy resolution of the monochromator is about 10^{-3}. Typical flux measured at the sample position at 17 keV is about 1.6 x 10^8 photons/mm^2.s with a stored electron beam of 300 mA as ELETTRA operates at 2 GeV [13]. A custom-built ionization chamber is placed upstream to the sample to determine the exposure on the sample. A micrometric vertical and horizontal translation stage allows the positioning and the scanning of the sample with respect to the stationary beam and a rotational stage allows CT acquisition with a resolution of 0.001°.

The detector system is comprised of a 16-bit CCD camera, with 2048 x 2048 pixels^2, 14 x 14 μm^2 pixel size, coupled to an intensifier screen with no magnification (1:1). The CCD camera is able to move along the sample-detector axis, in order to set the desired sample-to-detector distance *d*. According to the choice of the sample-to-detector distance, one may distinguish between the absorption and phase sensitive regimes. If the CCD is mounted very close to the sample we are in the absorption regime. For higher *d* values, free space propagation transforms the phase modulation of the transmitted beam into an amplitude modulation.

3. Artificial Neural Networks Training Strategies

The results presented are related to the usage of FF-ANNs with the training algorithms Gradient Descent with adaptive Learning Rate and Momentum Training (GDLRM), Resilient Backpropagation (RB), the quasi-newton BFGS, and Levenberg-Marquardt (LM) [18].

The combination of the training algorithms with different kernels and symmetry information yielded what we call training strategies. We were interested in presenting as input to the ANNs kernels ("masks" for the subimages) of different dimensions and symmetry information aiming to convey context information based on the pixels' intensity. We have tested the dimensions 3x3, 5x5 and 7x7 pixels in the training strategies.

In the assessment of training strategies we compared ways of conveying symmetry information, which we call *Asym*, *SymAv1* and *SymAv2*. In the conventional asymmetric approach (*Asym*), the numbers of inputs of the pixel's grayscale values were: 9 inputs for a 3x3 kernel, 25 inputs for a 5x5 kernel and 49 inputs for a 7x7 kernel. On the other hand, in the symmetric approach (*SymAv1* and *SymAv2*), the average of intensity (grayscale values) of different groups of pixels was used, as depicted in Figs. 2 and 3.

7	8	9	6	9	8	7
8	4	5	3	5	4	8
9	5	2	1	2	5	9
6	3	1	0	1	3	6
9	5	2	1	2	5	9
8	4	5	3	5	4	8
7	8	9	6	9	8	7

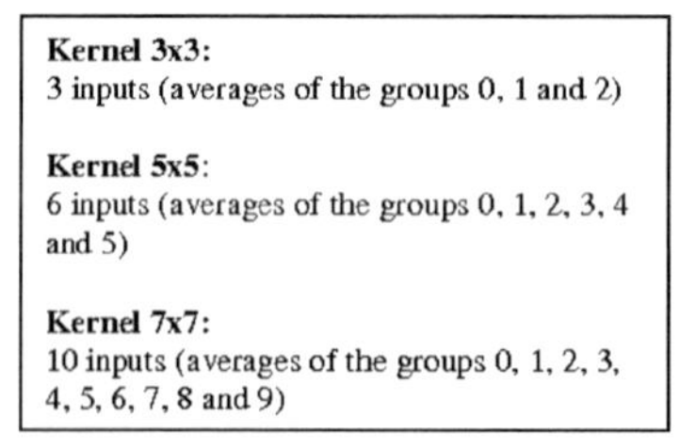

Figure 2. Symmetry information SymAv1 for the subimages used to train the ANNs (average intensity of groups of grayscale pixels, related to the orientation): 3 inputs for a 3x3 kernel – white pixels; 6 inputs for a 5x5 kernel – white and light gray pixels; and 10 inputs for a 7x7 kernel – white, light gray and dark gray pixels.

In the case of the average value for groups related to the orientation (*SymAv1*), Figure 2 exhibits 3 inputs for a *3x3* kernel (instead of 9 inputs), 6 inputs for a *5x5* kernel (instead of 25 inputs) and 10 inputs for a *7x7* kernel (instead of 49 inputs); for average values of groups surrounding the central pixel (*SymAv2*), Figure 3 exhibits 2 inputs for a *3x3* kernel (instead of 9 inputs), 3 inputs for a *5x5* kernel (instead of 25 inputs) and 4 inputs for a *7x7* kernel (instead of 49 inputs). The main idea with this approach is the following: with a smaller number of inputs, there are fewer weights to be adjusted in the ANN, reducing redundancy and thereby making ANN learning both more efficient and less prone to overfitting.

3	3	3	3	3	3	3
3	2	2	2	2	2	3
3	2	1	1	1	2	3
3	2	1	0	1	2	3
3	2	1	1	1	2	3
3	2	2	2	2	2	3
3	3	3	3	3	3	3

Kernel 3x3:
2 inputs (averages of the groups 0 and 1)

Kernel 5x5:
3 inputs (averages of the groups 0, 1 and 2)

Kernel 7x7:
4 inputs (averages of the groups 0, 1, 2 and 3)

Figure 3. Symmetry information SymAv2 for the subimages used to train the ANNs (average intensity of groups of grayscale pixels, surrounding the central pixel): 2 inputs for a 3x3 kernel – white pixels; 3 inputs for a 5x5 kernel – white and light gray pixels; and 4 inputs for a 7x7 kernel – white, light gray and dark gray pixels.

In short, each training algorithm (*GDLRM*, *RB*, *BFGS* and *LM*) was combined with a type of kernel (*3x3*, *5x5* or *7x7*) and one type of symmetry assumption (*Asym*, *SymAv1* or *SymAv2*), resulting in 9 combinations for each training algorithm, that is, 36 training strategies.

4. Validation Methodology: The Leave-One-Out Cross Validation

A *classifier* is a function which is able to map from unlabeled instances to class labels. It is induced by an algorithm that builds it from a data set [19]. In our case, the ANNs are the classifiers, and supervised learning (with the specificities of each training strategy as stated before) plays the role of the induction algorithm and the instances are the pixels to be classified. A description of the supervised learning methodology for ANNs is given in [6].

Some methods such as bootstrap, holdout or LOOCV may be used to estimate the future prediction accuracy of a classifier [19]. In our work we used the LOOCV and the Mean Square Errors provided by the LOOCV. The main idea with the LOOCV is that the classifier will be initialized and trained k times by the induction algorithm. Given a set S with k instances $S = \{X_1, X_2, \ldots, X_k\}$ and their respective classification, for each test t_i in $T = \{t_1, t_2, \ldots, t_k\}$, the classifier is trained with k-1 instances, since during the test t_i, the instance X_i is *left out* from the training phase in t_i, and X_i will be submitted to classification and therefore an estimation of misclassification of X_i can be calculated. In other words, the classifier is always trained with a subset of instances and tested with the instance that was not used in the training phase. Thus, it is possible to estimate the accuracy of classifiers based on the classification of an instance that was not presented during the training phase in order to compare, choose, combine or estimate the bias and variance of future processes of classification.

Our ANNs had one hidden layer with 5 neurons, and only one output for the classification (0 for no-bone, or 1 for bone). The number of neurons in the input layer depended on the training strategy, as mentioned in the section 3 (*Asym*: 9, 25 or 49 inputs; *SymAv1*: 3, 6 or 10 inputs; *SymAv2*: 2, 3 or 4 inputs).

In the present survey, 50 pixels of interest from the Figure 1 were selected and classified in order to compose the set *S* (25 bone pixels, 13 marrow pixels and 12 background pixels). The same set *S* was used for all training strategies, for all ANNs tested. During the training phases, the stopping criterion was either 4000 epochs or an MSE of the training phase lower than 10^{-5}. The pixel intensity information corresponding to the 49 instances, according to the information symmetries depicted in Figs. 2 and 3, was presented to the ANN as well as their classification during the training phase with a given training strategy. Then, the trained ANN was used for the classification of the *left-out* pixel, yielding, for the test t_i a square error SE_i of the output provided by the ANN in relation to the pixel's classification (0 for no-bone, or 1 for bone). For the test t_{i+1}, the ANN is re-initialized randomly and the process is started over again. Then, the MSE over the 50 SEs was calculated. This procedure (MSE for the LOOCV of the 50 pixels of the set *S*) was repeated 30 times for each training strategy, yielding 30 MSEs, and the box plots were plotted with the log of the MSEs for a better visualization.

Summarizing, each LOOCV yielded one MSE and we repeated 30 times the LOOCV for each training strategy. The comparison between the results is useful to determine which training strategies are likely to have a good accuracy in bone pixel classification for SR μCT images for HMM.

5. Computational Experimental Results

Images obtained by the segmentation of Figure 1 performed by ANNs (trained with strategies chosen after analyses) are presented in Figs. 4-7. They are illustrative examples of the final results. Although they look very similar, slight differences may be noticed. Figure 8 depicts the box plots (log MSE) of the best training strategies assumed for each training algorithm. Table 1 exhibits the average MSE, standard deviation and confidence interval of the mean at confidence level 95% for the best training strategies considered. Those and other results were reported in [20].

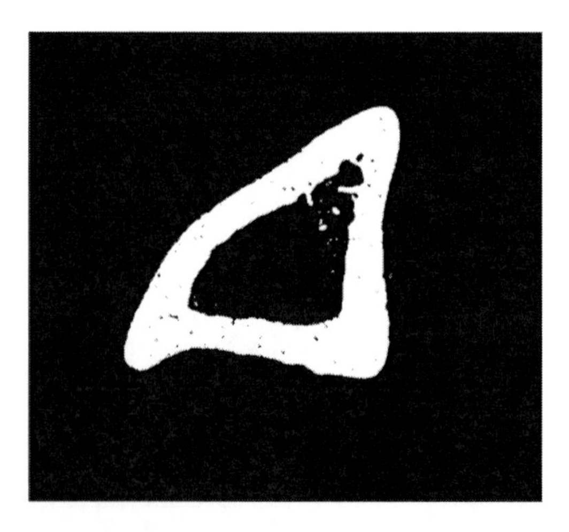

Figure 4. Segmentation of the image of interest (Figure 1) with the classification provided by the ANN trained with the strategy GDLRM 3x3 SymAv2.

Table 1. Average MSE, Standard Deviation on Interval of Confidence for the best training strategies considered

	Av. MSE	St. Dev.	t-Int1	t-Int2
GDLRM 3x3 SymAv2	2.69×10^{-3}	6.23×10^{-4}	2.46×10^{-3}	2.92×10^{-3}
RB 3x3 SymAv2	9.82×10^{-4}	5.99×10^{-4}	7.59×10^{-4}	1.21×10^{-3}
BFGS 3x3 SymAv1	1.52×10^{-4}	1.41×10^{-4}	9.90×10^{-5}	2.04×10^{-4}
LM 3x3 Asym	3.17×10^{-4}	9.37×10^{-4}	0	6.67×10^{-4}

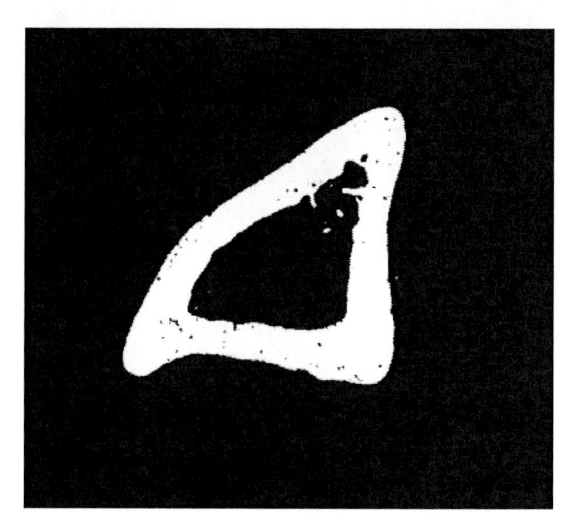

Figure 5. Segmentation of the image of interest (Figure 1) with the classification provided by the ANN trained with the strategy RB 3x3 SymAv2.

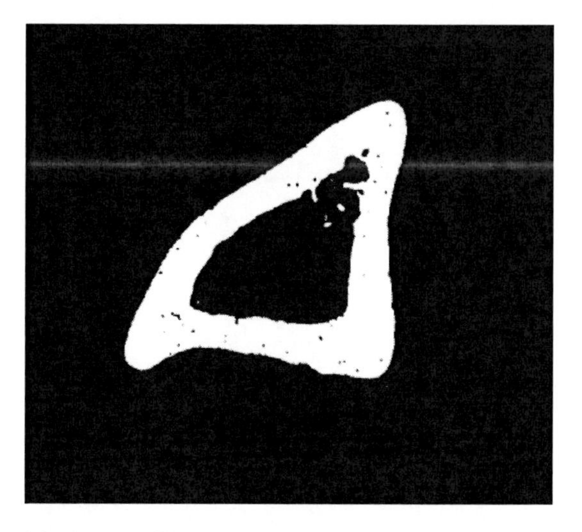

Figure 6. Segmentation of the image of interest (Figure 1) with the classification provided by the ANN trained with the strategy BFGS 3x3 SymAv1.

Figure 7. Segmentation of the image of interest (Figure 1) with the classification provided by the ANN trained with the strategy LM 3x3 Asym.

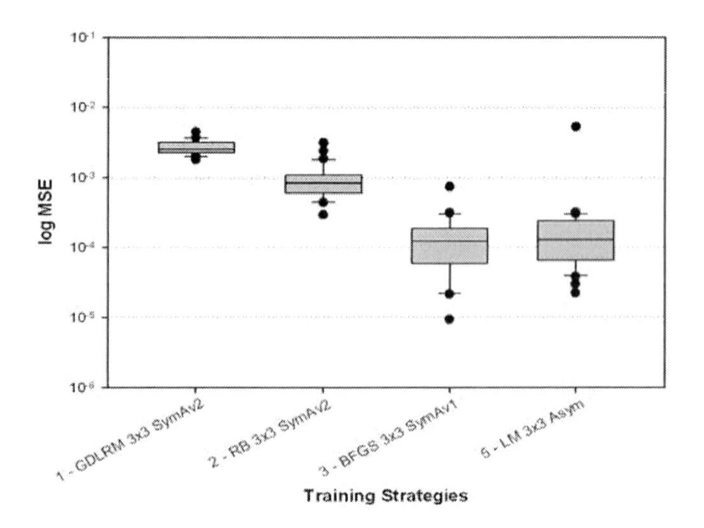

Figure 8. The boxplots of best training strategies considered for each training algorithm (GDLRM, RB, BFGS and LM).

6. Discussion

A visual analysis of the segmented images in Figs. 4-7 gives an idea of the classification processes for the image of interest (Figure 1), based on the classification task performed by the ANNs trained with the algorithms *GDLRM 3x3 SymAv2*, *RB 3x3 SymAv2*, *BFGS 3x3 SymAv1* and *LM 3x3 Asym*. The visual results of the segmentation are evidence that the training strategies presented are satisfactory. According to Table 1, it is possible to say that for all these combinations the average MSE is very low, but still lower for the best combinations of the algorithms *BFGS* and *LM*. Going back to the initial problem of recognizing bone pixels in SR-μCT images, our methodology of validation allows us to affirm that the tests were successful, in the sense that the ANNs achieve very low average MSE for the image of interest.

CONCLUSION

SR-μCT enables magnified images which may be used as a non-invasive and non-destructive biomedical imaging technique. However, as the μCT is relatively new, further investigation on AI algorithms and techniques are required. In this chapter, results of the assessment of well-known architectures and training algorithms for ANNs, symmetries and kernels were reported. Training strategies have yielded outstanding results for the bone pixel recognition, with a satisfactory segmentation of the image of interest based on the classification performed by the ANNs.

ACKNOWLEDGMENTS

The author A.A.M.M. would like to acknowledge the Brazilian agencies CAPES and FAPERJ for supporting this research (Grant E-26/102.493/2010). Portions of this chapter were published at the journal Nuclear Instruments and Methods in Physics Research Section A.

REFERENCES

[1] Gorry, GA. Computer-Assisted Clinical Decision Making. *Methods of Information in Medicine*, 1973, 12, 45-51.

[2] Clancey, WJ; Shortliffe, EH; Buchanan, BG. Intelligent Computer-Aided Instruction for Medical Diagnosis. *Proceedings of the Third Annual Symposium on Computer Applications in Medical Care*, USA, 1979.

[3] Szolovits, P; Patil, RS; Schwartz, WB. Artificial Intelligence in Medical Diagnosis. *Annals of Internal Medicine*, 1988, 108, 80-87.

[4] Brinkley, JF. Knowledge-Driven Ultrasonic Three-Dimensional Organ Modeling. *IEEE Transactions on Pattern Analysis and Machine Intelligence*, 1985, PAMI-7, 431-441.

[5] Brugal, G. Pattern recognition, image processing, related data analysis and expert systems integrated in medical microscopy. *Proceedings of the 9th International Conference on Pattern Recognition*, Italy, 1988.

[6] Haykin, S. *Neural Networks: a Comprehensive Foundation*. India: Pearson Education, 2005.

[7] Egbert, DD; Rhodes, EE; Goodman PH. Preprocessing of Biomedical Images for Neurocomputer Analysis. *Proceedings of the IEEE International Conference on Neural Networks*, 1998, 561-568.

[8] Egmont-Petersen, M; de Ridder, D; Handels, H. Image processing with neural networks - a review. *Pattern Recognition*, 2002, 35, 2279-2301.

[9] Hounsfield, GN. Computerized transverse axial scanning (tomography). *The British Journal of Radiology*, 1973, 46, 1016-1022.

[10] Elder, FR; Gurewitsch, AM; Langmuir, RV; Pollock, HC. Radiation from Electrons in a Synchrotron. *Physical Review*, 1947, 71, 829-830.

[11] Lewis, RA. Medical phase contrast x-ray imaging: current status and future prospects. *Physics in Medicine and Biology*, 2004, 49, 3573-3583.

[12] Olivo, A. Towards the exploitation of phase effects in synchrotron radiation radiology. *Nuclear Instruments and Methods in Physics Research Section A*, 2005, 548, 194-199.

[13] Arfelli, F *et al.* Mammography with Synchrotron Radiation Phase-Detection Techniques, *Radiology*, 2000, 215, 286-293.

[14] Chappard, D; Baslé, M-F; Legrand, E; Audran, M. Trabecular bone microarchitecture: a review. *Morphologie*, 2008, 92, 162-170.

[15] Betz, O; Wegst, U; Weide, D; Heethoff, M; Helfen, L; Lee, W-K; Cloetens, P. Imaging applications of synchrotron X-ray phase contrast microtomography in biological morphology and biomaterials science. I. General aspects of the technique and its advantages in the analysis of millimeter-size arthropod structure, *Journal of Microscopy*, 2007, 227, 51-71.

[16] Meneses, AAM; Pinheiro, CJG; Schirru, R; Barroso, RC; Braz, D; Oliveira, LF. Artificial Neural Networks Applied to Bone Recognition in X-Ray Computed Microtomography Imaging for Histomorphometric Analysis. *IEEE Nuclear Science Symposium and Medical Imaging Conference Record*, Germany, 2008, 5309-5313.

[17] Meneses, AAM; Pinheiro, CJG; Gambardella, LM; Schirru, R; Barroso, RC; Braz, D; Oliveira, LF. Neural Computing for Quantitative Analysis of Human Bone Trabecular Structures in Synchrotron Radiation X-Ray μCT Images. *IEEE Nuclear Science Symposium and Medical Imaging Conference Record*, USA, 2009, 3437-3441.

[18] Demuth, H; Beale, M. *Neural Network Toolbox*. The MathWorks, 2004.

[19] Kohavi, R. A Study of Cross-Validation and Bootstrap for Accuracy Estimation and Model Selection. *Proceedings of the International Joint Conference on Artificial Intelligence IJCAI*, USA, 1995, 1137–1143.

[20] Meneses, AAM; Pinheiro, CJG; Rancoita, P; Schaul, T; Gambardella, LM; Schirru, R; Barroso, RC; Oliveira, LF. Assessment of neural networks training strategies for histomorphometric analysis of synchrotron radiation medical images. *Nuclear Instruments and Methods in Physics Research Section A*, 2010, 621, 662-669.

INDEX

A

B

C

D

E

F

G

H

I

Q

R

S